PLURIPOTENT STEM CELL BIOLOGY - ADVANCES IN MECHANISMS, METHODS AND MODELS

Edited by **Craig S. Atwood and Sivan Vadakkadath Meethal**

Pluripotent Stem Cell Biology - Advances in Mechanisms, Methods and Models
http://dx.doi.org/10.5772/57025
Edited by Craig S. Atwood and Sivan Vadakkadath Meethal

Contributors

Jürgen Hescheler, Toshihiro Akaike, Kakon Nag, Hideyuki Kobayashi, Koichi Nagao, Koichi Nakajima, Toshihiro Tai, Naoki Nishishita, Minoru Tomizawa, Georges Lacaud, Monika Stefanska, Elias Zambidis, Jasmin Agarwal, Hiroshi Koide, Feng Ma, Bin Mao, Bo Chen, Ya Zhou, Yasuhiro Ebihara, Kohichiro Tsuji, Peter A. Clark

Published by InTech

Janeza Trdine 9, 51000 Rijeka, Croatia

Publishing Process Manager Iva Lipovic

Technical Editor InTech DTP team

Cover InTech Design team

First published June, 2014

Printed in Croatia

Additional hard copies can be obtained from orders@intechopen.com

Pluripotent Stem Cell Biology - Advances in Mechanisms, Methods and Models, Edited by Craig S. Atwood and Sivan Vadakkadath Meethal

p. cm.

ISBN 978-953-51-1590-8

For EU product safety concerns:
IN TECH d.o.o., Prolaz Marije Krucifikse Kozulić 3, 51000 Rijeka, Croatia,
info@intechopen.com or visit our website at intechopen.com.

Meet the editors

Craig S. Atwood, Ph.D., is an Associate Professor of Medicine at the University of Wisconsin and a Health Science Specialist with the Geriatric Research, Education and Clinical Center at the William S. Middleton Memorial Veterans Administration Hospital in Madison. Dr. Atwood completed his Ph.D. in Biochemistry from the University of Western Australia in Perth, Australia prior to post-doctoral fellowships at the National Cancer Institute, NIH, Bethesda, and Massachusetts General Hospital, Charlestown. He held faculty positions at Harvard Medical School and Case Western Reserve University prior to his current appointment where he directs the research program of the Laboratory for Endocrinology, Aging and Disease. Dr. Atwood has broad research interests related to the endocrinology of embryogenesis, adulthood and senescence as elaborated upon in 'The Reproductive-Cell Cycle Theory of Aging'. He has published over 200 scientific articles, has served on numerous review boards and is an Editor of 20 scientific journals including the Journal of Biological Chemistry. In 2006 he received the Zenith Fellows Award from the Alzheimer's Association in recognition of his research.

Sivan Vadakkadath Meethal, M.Sc., Ph.D., is a Senior Scientist in the Department of Medicine at the University of Wisconsin-Madison, Madison, Wisconsin, USA. Dr. Vadakkadath Meethal completed his Ph.D. in Biochemistry from the University of Kerala, Trivandrum, India prior to post-doctoral fellowships at the Indian Institute of Science, Bangalore, India and University of Wisconsin-Madison, Madison, Wisconsin, USA. Dr. Vadakkadath Meethal has research interests that focus on the endocrine and epigenetic regulation of neurogenesis, neurodegeneration, neuroregeneration, cancer and aging. He has published widely in these areas, is an editor of several scientific journals and has received numerous fellowships and awards for his research.

Contents

Preface

Great progress continues to be made in our understanding of stem cell self-renewal and pluripotency, and the utilization of stem cells for basic and applied applications. This book embodies recent advances in the biological mechanisms, methods and models of stem cell self-renewal, reprograming and regeneration, as well as touching on the ethical and moral dilemmas of embryo donation and adoption. In the first section of this book, 'Stem Cell Self-Renewal and Pluripotency', Agarwal and Zambidis examine the role of the NFκB-STAT3 signaling axis in regulating the induction and maintenance of the pluripotent state. The authors discuss a novel link between inflammatory pathways and efficient cell reprogramming, whereby bone marrow stromal-primed human myeloid cell progenitors are significantly more receptive to reprogramming stimuli than other cell types. Myeloid cells appear to harbor a unique epigenetic plasticity and are innately equipped to transcriptionally and epigenetically activate key inflammatory pathways via an interconnected NFκB and STAT3 signaling machinery. Both pathways act as epigenetic modifiers which promote ESC pluripotency by inducing an open chromatin state that allows transcription factors to regulate cell fates. The importance of the NuRD complex in ensuring that differentiated cells do not reactivate pluripotency genes, which might enable tumorigenesis – is also discussed in this first chapter. The following chapter by Koide examines further the idea that many genes involved in ESC self-renewal also are involved in cancer cell growth. Considerable evidence supports the author's contentions, where self-renewal genes such as Oct 3/4, Sox2, Nanog, STAT3, Klf4 and Zfp57 are not only required for self-renewal in ESCs, but also are highly expressed in neoplastic cells. Conversely, there is evidence that oncogenes also are expressed in SCs. That several common transcription factors are involved in the regulation of ESC self-renewal and cancer cell growth raises the intriguing possibility that these common transcription factors are specifically expressed in cancer stem cells involved in tumor growth. Moreover, the authors suggest that since ESCs and induced pluripotent stem cells (iPSCs) have similar gene expression profiles, tumor risk also might be elevated for iPSCs used for therapeutic purposes. As suggested by Agarwal and Zambidis, preventing cancerous epigenetic patterns in iPSC via more accurate high-fidelity reprogramming methods will be the foundation for future clinical applications.

In the next section on 'Haematopoiesis', Stefanska and colleagues review how ESCs have been used to study the development of the haematopoietic system, that is otherwise very difficult to study in vivo. As the authors write, 'ESCs have been instrumental in identifying and characterizing the elusive haemangioblast .... and more recently, this model system has allowed the merging of two conflicting theories of the origin of blood cells (haemangioblast and haemogenic endothelium) into a single linear model of development'. In addition, the precise roles and requirements of many critical regulators of this process have been elucidated using this model system. This system offers many opportunities to further study cell signaling pathways that support the development of normal haematopoiesis and leukaemogenesis. Compared with the murine haematopoietic system, however, little is known regarding the molecular and cellular regulation of early hematopoiesis in the human as pointed out in the next review by Chen and colleagues. These researchers have established an efficient method to induce large-scale production of multipotential hematopoietic progenitor cells by co-culturing hESC/hiPSCs with murine hematopoietic niche-derived stromal cells. Their review discusses the origin, evolution and the development of both primitive and definitive hematopoietic waves, especially those derived from hESCs in vitro systems. They also summarize the cellular and molecular characteristics of cells in primitive and/or definitive hematopoiesis as well as the critical problems and challenges facing scientists working in this important area of research. With the advent of these novel hESC/hiPSCs, our understanding of human haematopoietic development should proceed swiftly.

Major discoveries in the regulation of self-renewal, pluripotency and differentiation are dependent upon technical developments in the field. In this next section on the 'Technical Advances in the Culture and Use of Induced Pluripotent and Embryonic Stem Cells', three research groups present advances and applications of hESCs/iPSCs. Nishishita and colleagues demonstrate a robust, low-cost, stable method for generating and maintaining iPSC clones from cord blood (CB) cells. This feeder-free and serum-free method that utilizes a temperature sensitive Sendai virus vector also solves some of the safety concerns related to tumorigenicity arising from chromosomal integration of exogenous genes and/or infectious hazards associated with the use of by xenogeneic biological products in the culture system. The use of CB cells, the youngest somatic cells, is suggested to alleviate concerns regarding post-natal DNA damage and the ability to cryopreserve CB HSCs long-term in bank confers a unique advantage to CB cells as a suitable material for generating induced pluripotent stem (iPSC) cells for future clinical use. In the next chapter, Nag and colleagues review their development of cell recognizable Fc-chimeric proteins aimed at providing simultaneous support for cell adhesion, proliferation and differentiation as a highly effective novel class of defined biomaterials for stem cell applications. These biomaterials have application in regenerative medicine and tissue engineering, including feeder-cell free ESC culture, a simplified and cost-effective culture system for stem cells, directed differentiation of stem cells, and on-site stress-free purification of target cells. In the final chapter of this section, Tomizawa and colleagues describe the development of a hepatocyte selection media that can be used to select hepatoblast-like cells from ESC and other cells. This clever approach utilizes differences in arginine and glucose metabolism between SCs and differentiated hepatocytes to eliminate (induce death of) hiPSCs, for example, and the isolation of differentiated hepatocytes. This methodology avoids the damage introduced by other selection protocols and is an important contribution to the development of pure hepatocytes for culture or applied applications.

The next two chapters examine 'New Stem Cell Models' in two divergent systems. The first, by Potta and colleagues reviews the potential of using stem cells for evaluating the safety of drugs. Drug discovery programs often utilize animals to test the efficacy and safety of new drugs, but the results from such experiments cannot always be extrapolated to humans. This chapter reviews the development of iPSCs as a novel and cost-effective source of organotypic cells for assessing drug toxicity. Kobayashi and colleagues review the use and challenges of using domestic pigs and the minipig to both derive and utilize SCs.

In the final section on 'Ethics', Clark and colleagues remind us of the medical, legal and ethical dilemmas associated with embryo donation/adoption. They contend that allowing for embryo donation/adoption is the only viable option that protects and preserves their human life. The other viable options: being discarded, destroyed for research, abandoned or kept in "suspended animation" indefinitely, are unacceptable because they have the potential of harming or intentionally killing these embryos that deserve special respect. To avoid these later situations, Clark puts forth a number of recommendations and safeguards involving: the matching of eggs to be fertilized with those placed in the uterus of the mother, the enactment of laws at the federal level that regulates Assisted Reproductive Technologies and that regulates the creation, destruction and exploitation of human embryos, that infertile couples and individuals willing to take full responsibility for the upbringing of these children should be encouraged to consider adoption of the presently existing frozen embryos, and that children who are adopted from frozen embryos have the right to know their genetic make-up.

The reviews in this book are a reminder of the rapid progress being made in our understanding of stem cell self-renewal and pluripotency, the methodological advances in the culture, purification and use of stem cells, and how this basic knowledge and methodological advances can be utilized for future regenerative medicine, drug screening and other applications of medical benefit.

Craig S. Atwood and Sivan Vadakkadath Meethal
Geriatric Research, Education and Clinical Center,
Veterans Administration Hospital,
Department of Medicine,
University of Wisconsin,
Madison, USA

Section 1

Stem Cell Self-Renewal and Pluripotency

Chapter 1

The Role of an NFκB-STAT3 Signaling Axis in Regulating the Induction and Maintenance of the Pluripotent State

Jasmin Roya Agarwal and Elias T. Zambidis

Additional information is available at the end of the chapter

http://dx.doi.org/10.5772/57602

1. Introduction

Induced pluripotent stem cells (iPSC) are generated by reprogramming differentiated somatic cells to a pluripotent cell state that highly resembles embryonic stem cells (ESC) [1]. Fully reprogrammed iPSC can differentiate into any adult cell type [2-6]. Takahashi and Yamanaka generated the first iPSC in 2006 by transfecting fibroblasts with four defined factors: SOX2, OCT4, KLF4, c-MYC (SOKM; also referred to as Yamanaka factors) [7]. The clinical use of iPSC offers great potential for regenerative medicine as any cell type can be generated from true pluripotent cells [8-10]. However, human clinical iPSC applications are currently limited by inefficient methods of reprogramming that often generate incompletely reprogrammed pluripotent states that harbor potentially cancerous epigenetic signatures, and possess limited or skewed differentiation capacities [11-13]. Many standard iPSC lines do not fully resemble pluripotent ESC, and often retain an epigenetic memory of their cell of origin [14, 15]. Such incompletely reprogrammed iPSC also display limited differentiation potential to all three germ layers (e.g., endoderm, ectoderm, mesoderm) [16, 17].

To avoid integrating retroviral constructs that may carry mutagenic risks, many non-viral methods have been described for hiPSC derivation [18, 19]. For example, one successful approach is to transiently express reprogramming factors with EBNA1-based episomal vectors [20-22]. It was initially intuitive to reprogram skin fibroblasts due to their easy accessibility. However, standard episomal reprogramming in fibroblasts occurs at even lower efficiencies (< 0.001-0.1%) than reprogramming with retroviral vectors (0.1%–1%) [23-25]. Subsequent studies revealed that various cell types possess differential receptiveness for being reprogrammed to pluripotency [26-30]. One highly accessible human donor source is blood, which has been demonstrated to reprogram with significantly greater efficiency than fibroblasts [4, 20, 31-33].

The innate immune system possesses highly flexible cell types that are able to adapt quickly to various pathogens by eliciting defense responses that protect the host [34-36]. Innate immune cells derived from the myeloid lineage (eg, monocyte-macrophage, dendritic cells, neutrophils) are able to reactivate some unique features of pluripotent stem cells that may give them greater flexibility for being reprogrammed to a pluripotent cell state than other differentiated cells [37]. Additionally, the differentiation state of the cell seems to be of critical importance for its reprogramming efficiency [38].

Our group established a reprogramming method that solves many of the technical caveats cited above (Figure 1). We have generated high-fidelity human iPSC (hiPSC) from stromal-primed (sp) myeloid progenitors [20]. This system can reprogram >50% of episome-expressing myeloid cells to high-quality hiPSC characterized by minimal retention of hematopoietic-specific epigenetic memory and a molecular signature that is indistinguishable from bona fide human ESC (hESC). The use of bone marrow-, peripheral-or cord blood (CB)-derived myeloid progenitor cells instead of fibroblasts, and a brief priming step on human bone marrow stromal cells / mesenchymal stem cells (MSC) appeared to be critical for this augmented reprogramming efficiency. In this system, $CD34^+$ - enriched cord blood cells (CB) are expanded with the growth factors (GF) FLT3L (FMS-like tyrosine kinase 3 ligand), SCF (stem cell factor) and TPO (thrombopoietin) for 3 days, subsequently nucleofected with non-integrating episomes expressing the Yamanaka factors (4F, SOX2, OCT4, KLF4, c-MYC), and then co-cultured on irradiated MSC for an additional 3 days. Cells are then harvested, and passaged onto MEF (mouse embryonic fibroblasts), and hiPSC are generated via standard methods and culture medium. The initial population of enriched $CD34^+$ CB progenitors quickly differentiates to myeloid and monocytic cells in this system, and reprogrammed cells arise from $CD34^-$ myeloid cells. The first iPSC colonies appear around day 10, and stable mature iPSC colonies can be established after ~21-25 days. The episomal constructs are partitioned after relatively few cell divisions (e.g., 2-9 passages) to generate high quality non-integrated hiPSC.

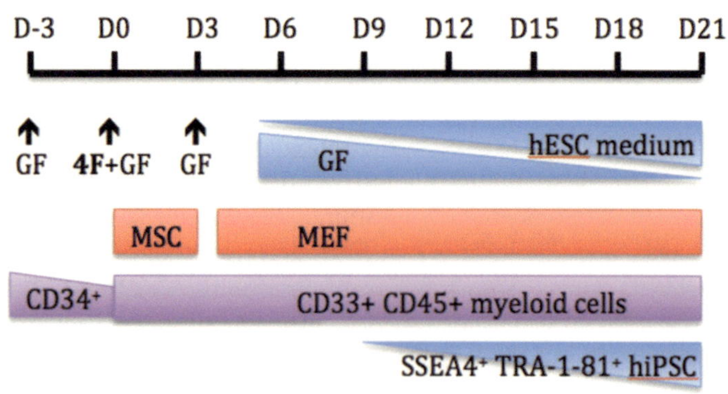

Figure 1. Schema of the stromal-primed myeloid reprogramming protocol for the generation of high quality human iPSC. 4F: four Yamanaka factors, GF: hematopoietic growth factors.

A proteomics and bioinformatics analysis of this reprogramming system implicated significant activation of MSC-induced inflammatory TLR-NFκB and STAT3 signaling [20]. A combination of cell contact-dependent and soluble factors mediate these effects. A recent study similarly implicated inflammatory TLR3 signaling as a novel trigger for enhanced fibroblast reprogramming, albeit at much lesser efficiencies than observed in our myeloid reprogramming system. TLR3 signaling leads to epigenetic modifications that favor an open chromatin state, which increases cell plasticity and the induction of pluripotency [39]. Lee *et al.* termed this novel link between inflammatory pathways and cell reprogramming 'Transflammation' [40].

In this chapter we will discuss hypotheses why inflammation-activated myeloid cells may be highly receptive to factor-mediated reprogramming. Specifically, we will explore the role of the NFκB-STAT3 signaling axis in mediating the unique susceptibility of myeloid cells to high-quality human iPSC derivation.

2. Overview of the canonical and non-canonical NFκB pathway

Multipotent myeloid progenitors are derived from hematopoietic stem cells and differentiate to monocytes macrophages, dendritic cells, and granulocytes, which elicit the initial innate immune response toward pathogens [41]. NFκB (nuclear factor kappa-light-chain-enhancer of activated B cells) is a central transcription factor that regulates these innate immune responses during microbial infections [42-44]. The NFκB system belongs to a group of early-acting transcription factors that are present in the cytoplasm in an inactive state but can be quickly activated by multiple inflammatory stimuli [45, 46].

2.1. The canonical NFκB signaling pathway

The NFκB family consists of 5 members; p65 (RelA), p50 and c-Rel are involved in canonical signaling, and p52 and RelB are involved in non-canonical signaling. Canonical NFκB signaling is characterized by activation of the IκB kinase complex (IKK), which contains two kinases, IKK1/α and IKK2/β along with a non-catalytic subunit called IKKγ (NEMO) [47, 48]. Unstimulated NFκB is sequestered in the cytoplasm by IκBα protein. In contrast, activation of the IKK complex (e.g., by TLRs) leads to IKKβ-mediated serine phosphorylation of IκBα triggering its proteasome-mediated degradation and its dissociation from NFκB [49, 50]. This activates the p65:p50 dimer through p65 phosphorylation and leads to NFκB translocation into the nucleus where it induces target gene expression. Subsequent acetylation keeps p65 in the nucleus [51]. This can be reverted by HDAC3 (histone deacetylase 3)-induced deacetylation of p65, which increases the affinity of NFκB proteins for IκBα and nuclear export [52, 53]. Canonical NFκB signaling is a fast and transient process that regulates complex inflammatory processes that includes the initial pro-inflammatory phase, the induction of apoptosis, and even tumorigenesis [54]. It can be activated by toll-like receptors (TLR), which recognize characteristic pathogenic molecules to activate innate immune responses [55-57].

2.2. The non-canonical NFκB signaling pathway

Non-canonical NFκB signaling is stimulated via the NFκB-inducing kinase (NIK), which leads to phosphorylation of the p100 precursor protein and generation of the p52:RelB dimer that translocates to the nucleus to activate gene transcription. This pathway is uniquely dependent on steady state levels of *NIK* expression, which are controlled under normal conditions through TRAF3-directed ubiquitination and proteasomal degradation. Non-canonical NFκB signaling is slow but persistent and requires de novo NIK protein synthesis and NIK stabilization [58]. It is activated by receptors that belong to the TNFR (tumor necrosis factor receptor) superfamily like BAFF (B-cell-activating factor), CD40 or lymphotoxin β-receptor (LTβR) [59-62].

The common feature of these receptors is the possession of a TRAF-binding motif, which recruits TRAF members (e.g., TRAF2 and TRAF3) during ligand ligation [63, 64]. Receptor recruitment of TRAF members triggers their degradation, and leads to NIK activation and p100 processing [65]. Additionally, BAFF is an important component of pluripotency-supporting growth media for the culture of ESC and a regulator of B-cell maturation [66]. It predominantly activates non-canonical NFκB signaling due to its possession of an atypical TRAF-binding sequence, which interacts only with TRAF3 but not with TRAF2 [67]. TRAF3 degradation is sufficient to trigger non-canonical NFκB signaling, whereby activation of the canonical NFκB pathway requires TRAF2 recruitment [68].

2.3. CD40 stimulates both NFκB pathway components

Another receptor associated with NFκB signaling is CD40, which is expressed on various cell types including B cells and monocytes. The CD40 receptor interacts with its ligand CD40L, which is primarily expressed on activated T cells. This signaling is majorly involved in B-cell activation, dendritic cell maturation, antigen presentation and acts as a co-stimulatory pathway of T-cells [69]. Upon ligation by CD40L, CD40 targets both the canonical and non-canonical NFκB pathways via proteolysis of TRAF2 and TRAF3 [70-72]. Non-canonical NFκB signaling regulates hematopoietic stem cell self-renewal via regulating their interactions with the microenvironment [73]. The deregulation of non-canonical hematopoietic NFκB signaling is associated with auto-immunity, inflammation and lymphoid malignancies [58, 74].

2.4. NFκB subunit functions

A third NFκB signaling pathway is activated following response to DNA damage that results in IκB degradation independent of IKK. This results in dimerization of free NFκB subunits that are mobilized similarly to canonical NFκB signaling [47]. Unlike RelA, RelB, and c-Rel, the p50 and p52 NFκB subunits do not contain transactivation domains in their C-terminus. Nevertheless, the p50 and p52 NFκB members play critical roles in modulating the specificity of NFκB functions and form heterodimers with RelA, RelB, or c-Rel [75]. Cell contact-dependent signals are crucial during immune responses and can be mediated through NFκB signaling [76]. This can be augmented by co-stimulatory signals like CD40 or CD28 that directly bind to NFκB proteins like p65 [77-81].

3. Functional role of NFκB signaling in stem cells

3.1. Differential roles of canonical and non-canonical NFκB signaling in embryonic stem cells

TLR activation is not only important for mediating innate immune responses, but also for stem cell differentiation. For example, hESC are characterized by the expression of pluripotency genes and markers such as OCT4, NANOG, alkaline phosphatase (AP) and telomerase [82-86]. NFκB signaling has been demonstrated to be crucial for maintaining ESC pluripotency and viability, and drives lineage-specific differentiation [87, 88]. A balance of canonical and non-canonical NFκB signaling regulates these opposing functions; non-canonical pathway signaling maintains hESC pluripotency, and canonical pathway signaling regulates hESC viability and differentiation [89, 90]. For example, non-canonical NFκB signaling has to be silenced during cell differentiation, which allows this pathway to act like a switch between hESC self-renewal and differentiation. RelB positively regulates several key pluripotency markers and represses lineage markers by direct binding to their regulatory units. RelB down-regulation reduces the expression of pluripotency genes like *SOX2* and induces differentiation-associated genes like *BRACHYURY* (mesodermal marker), *CDX2* (trophoectodermal marker) and *GATA6* (endodermal marker) [89].

3.2. Canonical NFκB signaling in hematopoietic stem cells

RelB/p52 signaling also positively regulates hematopoietic stem-progenitor cell (HSPC) self-renewal in response to cytokines (e.g., TPO and SCF) and maintains osteoblast niches and the bone marrow stromal cell population. It negatively regulates HSPC lineage commitment through cytokine down-regulation in the bone marrow microenvironment, although it is able to direct early HSC commitment to the myeloid lineage [73, 91].

Canonical p65 signaling also regulates hematopoietic stem cell functions and lineage commitment by controlling key factors involved in hematopoietic cell fate [92-94]. Canonical NFκB signaling is positively regulated by Notch1, which facilitates nuclear retention of NFκB proteins and promotes self-renewal [95-98]. FGF2 (fibroblast growth factor 2) is important for hESC self-renewal and preserves the long-term repopulating ability of HSPC through NFκB activation [99-102]. Deletion of p65, p52 and RelB dramatically decreases HSC differentiation, function and leads to extramedullary hematopoiesis [103]. NFκB pathway components and FGF4 are highly expressed in $CD34^{+}$HSPC from cord blood, where they regulate clonogenicity. Nuclear p65 can be detected in 90% CB-derived $CD34^{+}$ cells but only in 50% BM-derived $CD34^{+}$ cells [104]. The important role of NFκB in regulating myeloid cell lineage development has been most potently revealed via genetic deletion of IKKβ, IκBα, and RelB, which resulted in granulocytosis, splenomegaly and impaired immune responses [73, 103].

3.3. Canonical NFκB signaling during ESC differentiation

Canonical NFκB signaling is very low in the undifferentiated pluripotent state, where it maintains hESC viability. However, it strongly increases during lineage-specific differentiation of pluripotent stem cells. p65 binds to the regulatory regions of similar differentiation genes as RelB with opposing effects on their activation or silencing. It regulates cell proliferation by direct binding to the CYCLIN D1 promoter [89]. There are different levels of inhibiting canonical NFκB signaling: first, p65 translational repression by the microRNA cluster miR-290 to maintain low p65 protein amounts and second, the inhibition of translated p65 by physical interaction with NANOG. Similarly, *OCT4* expression is reversely correlated with canonical NFκB signaling [105]. In contrast to most observations in mouse ESC, NFκB probably plays a more important role in the maintenance of human ESC pluripotency [106]. Finally, active TLRs are expressed on embryonic, hematopoietic and mesenchymal stem cells (MSC), thus implicating their roles in a variety of stem cell types [107-110].

4. Role of NFκB signaling during reprogramming to pluripotency

Undifferentiated human iPSC have elevated NFκB activities, which play important roles in maintaining *OCT4* and *NANOG* expression in pluripotent hiPSC [111]. Innate immune TLR signaling was recently shown to enhance nuclear reprogramming probably through the induction of an open chromatin state, and global changes of epigenetic modifiers [39]. This normally increases cell plasticity in response to a pathogen, but may also enhance the induction of pluripotency, transdifferentiation and even malignant transformation [112-116].

The EBNA (Epstein-Barr virus nuclear antigen) is a virus-derived protein that is not only a critical component of episomal reprogramming vectors, where it mediates extra-chromosomal self-replication, but it is also known to activate several TLRs [117-119]. These include TLR3, which is known to augment reprogramming efficiencies through the activation of inflammatory pathways [39, 120]. TLR3 recognizes double-stranded RNA from retroviruses and signals through TRAF6 and NFκB [121-123]. The TLR3 agonist poly I:C was shown to have the same effect as retroviral particles in enhancing Yamanaka factor-induced iPSC production. TLR3 causes widespread changes in the expression of epigenetic modifiers and facilitates nuclear reprogramming by inducing an open chromatin state through down-regulation of histone deacetylases (HDACs) and H3K4 (histone H3 at lysine 4) trimethylations [38, 39, 124]. These epigenetic modifications mark transcriptionally active genes, whereas the H3K9me3 (Histone H3 at lysine 9) modification marks transcriptionally silenced genes [125, 126]. Histone deacetylation is generally associated with a closed chromatin state and HDAC inhibitors were shown to enhance nuclear reprogramming [127, 128]. Histone acetylation favors an open chromatin state, and is maintained by proteins containing histone acetyltransferase (HAT) domains, such as p300 and CBP [129, 130]. Interestingly, p300/CBP is able to interact with NFκB [131, 132]. RelB directly interacts with the methyltransferase G9a to mediate gene silencing of differentiation genes [133]. Epigenetic changes that allow an open chromatin state are crucial for giving the Yamanaka factors access to promoter regions necessary for the induction of pluripotency. Epigenetic chromatin modifications by TLRs are normally involved in the expression of host defense genes during infections [134-136]. This capability can be deployed to enable nuclear reprogramming as TLR3 was shown to change the methylation status of the *Oct4* and *Sox2* promoters. Interestingly, changes in these methylation marks were not observed with TLR3 activation alone but only in the presence of the reprogramming factors. Although TLR3 by itself promotes an open chromatin configuration, the reprogramming proteins are likely necessary to direct the epigenetic modifiers to the appropriate promoter sequences [137]. Lee *et al.* described the potential of inflammatory pathways to facilitate the induction of pluripotency as 'transflammation' [40, 138].

5. Overview of the JAK/STAT pathway

The JAK/STAT pathway (Janus kinase/signal transducer and activator of transcription) integrates a complex network of exterior signals into the cell, and can be activated by a variety of ligands and their receptors [139]. These receptors are associated with a JAK tyrosine kinase at their cytoplasmic domain. The JAK family consists of the four members JAK1, JAK2, JAK3 and TYK2 [140, 141]. Many cytokines and growth factors signal through this pathway to regulate immune responses, cell proliferation, differentiation and apoptosis [142-146]. Ligand binding induces the multimerization of gp130 receptor subunits, which brings two JAKs close to each other inducing trans-phosphorylation. Such activated JAKs phosphorylate their receptor at the C-terminus and the transcription factor STAT at tyrosine residues. This allows STAT dimerization and their nuclear translocation to induce target gene transcription. [147, 148] STAT3 acetylation is critical for stable dimer formation and DNA binding [149]. From the 7 mammalian STATs, STAT3 and STAT5 are expressed in many cell types, are activated by a plethora of cytokines and growth factors, and integrate complex biological signals [150, 151]. The other STAT proteins mainly play specific roles in the immune response to bacterial and viral infections. STAT3 is an acute phase protein with important functions during immediate immune reactions [152-154]. STAT3 can be recruited by receptor tyrosine kinases that harbor a common STAT3 binding motif in their cytoplasmic domain (e.g., GCSF (granulocyte colony-stimulating factor), LIF (leukemia inhibitory factor), EGF (epidermal growth factor), PDGF (platelet-derived growth factor), interferons (IFNγ) and interleukins (IL-6, IL-10)) [155-158]. Many cytokines signal through IL-10/STAT3 to achieve an immunosuppressive function or anti-apoptotic effect [159, 160]. IL-10 is also required during terminal differentiation of immunoglobulins [161]. STAT3 can be phosphorylated at tyrosine or serine residues. The phosphorylation site can play distinct roles in the regulation of downstream gene transcription [162]. Stat3-deficient mice die during early embryogenesis due to Stat3 requirement for the self-renewal of ESC [163].

Negative feedback regulation of the JAK/STAT circuitry is mediated by the SOCS family of target genes (suppressors of cytokine signaling) in a way that activated STAT induces SOCS transcription [164, 165]. SOCS proteins can bind to phosphorylated JAKs as a pseudo-substrate to inhibit JAK kinase activity and turn off the pathway [166, 167]. SOCS are negative regulators of the immune response [168, 169]. A small peptide antagonist of SOCS1 was shown to bind to the activation loop of JAK2 leading to constitutive STAT activation and TLR3 induction. This boosts the immune system to exert broad antiviral activities [170]. The JAK/STAT pathway also interacts with many other signaling pathways in a complex manner to regulate cell homeostasis and immune reactions [149, 171].

6. Functional role of the JAK/STAT pathway in stem cells

6.1. Stat3 maintains naïve pluripotency in mouse embryonic stem cells

ESC pluripotency is regulated by transcriptional networks that maintain self-renewal and inhibit differentiation [172-174]. Stat3 and Myc are necessary to maintain mouse ESC (mESC) self-renewal and bind to many ESC-enriched genes [175]. Their target genes include pluripotency-related transcription factors, polycomb group repressive proteins, and histone modifiers [176, 177]. The transcription factor Stat3 is a key pluripotency factor required for ESC self-renewal [178, 179]. Mouse ESC require LIF-Stat3 (leukemia inhibitory factor) and Bmp4 (bone morphogenic protein 4) to remain pluripotent in *in vitro* cultures, whereas human ESC require FGF2/MAPK (fibroblast growth factor / mitogen-activated protein kinase) and TGFβ/Activin/Nodal (transforming growth factor β) [180-183]. Nevertheless, the core circuitry of pluripotency is conserved among species and includes OCT4, SOX2 and NANOG [174].

6.2. The LIF-IL6-STAT3 circuitry

LIF belongs to the IL-6 family of cytokines and acts in parallel through the Jak/Stat3 and PI3K/Akt (Phosphatidylinositide 3-kinase) pathways to maintain *Oct4, Sox2* and *Nanog* expression via Kruppel-like factor 4 (Klf4) and T-box factor 3 [184, 185]. Lif and IL-6 are necessary for STAT3 phosphorylation mediated by Jak1 [186]. Stat3 phosphorylation positively regulates *Klf4* and *Nanog* transcripts and facilitates Lif-dependent maintenance of pluripotency in a signaling loop [106]. Stat3 directly binds to genomic sites of *Oct4* and *Nanog*, regulates the Oct4-Nanog circuitry and is necessary to maintain the self-renewal and pluripotency of mESC [187-189]. Overexpression of *Stat3* maintains mESC self-renewal even in the absence of Lif [190]. Withdrawal of LIF up-regulates the NFκB pathway and results in ESC differentiation as well as STAT3 disruption [191-193]. The interleukin 6 (IL-6) response element (IRE) is activated by STAT3, vice versa IL-6 stimulation leads to STAT3 phosphorylation and transactivation of IRE- containing promoters providing a positively regulated STAT3-IL6 loop. STAT3 directly associates with c-Jun and c-Fos in response to IL-6 [194]. c-Jun and c-Fos are DNA binding proteins and components of the AP-1 (activation protein-1) transcription factor complex [195]. AP-1 can be activated by TLR2/4, IL-10 or STAT3 to regulate inflammatory responses or drive keratinocyte differentiation in interplay with STAT3 and c-MYC [196]. Tlr2 also plays an important role in the maintenance of mESC [107]. STAT3 is important to tune appropriate amounts of AP-1 proteins required for proper differentiation. DNA binding sites for both *AP-1* and *STAT3* have been found in many gene promoters [194, 197]. It is important to note that c-Jun is able to capture or release the NuRD (nucleosome remodeling and deacetylation) repressor complex, an important epigenetic modulator of gene silencing [198, 199]. STAT3 is able to bind to bivalent histone modifications enabling a quick switch between the activation of pluripotency genes during ESC maintenance and their inhibition during cell differentiation [193].

6.3. STAT3 signaling in immune cells

STAT3 also has complex functions during hematopoietic development, immune regulation, cell growth, and leukemic transformation [200-202]. It is critically important for the survival and differentiation of lymphocytes and myeloid progenitors [171]. STAT3 signaling can be activated in a cell contact-dependent way, which is distinct from its cytokine activation. Co-cultures of MSC (human mesenchymal stem cells) and APC (antigen-presenting cell) increase STAT3 signaling in both cell types in a cell contact-dependent way, which mediates the immune-modulatory effects of MSC to block APC maturation and induce T-cell tolerance [203]. MSC are high-proliferative non-hematopoietic stem cells with the ability to differentiate into multiple mesenchymal lineages [204-206]. They accumulate in tumor environments in response to NFκB signaling and produce cytokines [207]. MSCs are FDA-approved for the treatment of severe acute GVHD, due to their immunomodulatory properties [208]. STAT3 phosphorylation is induced by cell-cell contacts and inhibited in postconfluent cells that consequently become apoptotic. Therefore, STAT3 may represent a molecular junction that allows cell proliferation or growth arrest depending on the state of the cell. Increased STAT3 activity may promote cell survival during cell confluency [209].

6.4. Cell contact-dependent STAT3 signaling during cell transformation

Constitutive STAT3 activation can by itself result in cellular transformation [210-214]. For example, contact-dependent STAT3 activation is known to play a promoting role in the interactions between tumor cells and their environment [215-218]. Cell transformation and the induction of pluripotency may share very similar signaling processes, and it is possible that STAT3 may represent a common axis [219, 220]. During early tumor development, certain cells have to acquire stem cell-like features that allow them to self-renew (tumor-initiating cells) and to produce cell progeny (tumor bulk) [221-224]. These tumor-initiating cells are very difficult to eradicate during chemotherapies and often re-establish the tumor seen as clinical relapse [225-227]. Tumor-initiating cells display strong inflammatory gene signatures with elevated IL6-STAT3-NFκB signaling to sustain their self-renewal [228-231]. A better understanding of the mechanism by which STAT3 and NFκB regulates the acquisition of pluripotency and self-renewal might also give us crucial insight about tumor development, and may lead to future novel therapies [171, 232].

7. The role of STAT3 signaling during reprogramming

7.1. STAT3 is a master reprogramming factor

Activation of Stat3 is a limiting factor for the induction of pluripotency, and its over-expression eliminates the requirement for additional factors to establish pluripotency [233]. These key properties have positioned Stat3 signaling as one of the master reprogramming factors that dominantly instructs naïve pluripotency [175]. Elevated Stat3 activity overcomes the pre-iPSC reprogramming block and enhances the establishment of pluripotency induced by SOKM [234]. Stat3 and Klf4 co-occupy genomic sites of *Oct4, Sox2* and *Nanog*. Klf4 and c-Myc are downstream targets of Stat3 signaling and part of the transcriptional network governing pluripotency. The Stat3 effect is combinatorial with other reprogramming factors, which implies that additional targets of Stat3 play a pivotal role [235].

7.2. STAT3 is an epigenetic regulator

Stat3 activation regulates major epigenetic events that induce an open-chromatin state during late-stage reprogramming to establish pluripotency [236-238]. For example, Stat3 signaling stimulates DNA methylations to silence lineage commitment genes and facilitates DNA demethylations to activate pluripotency-related genes [106, 239, 240]. Other chromatin modifications include histone acetylation and deacetylation, which are catalyzed by enzymes with histone acetyltransferase (HAT) or histone deacetylase (HDAC) activities. Histone acetylation is associated with an open chromatin state that allows active gene transcription. HDAC inhibitors are known to significantly improve the efficiency of iPSC generation by allowing promoter accessibility [128, 241, 242]. STAT3 suppresses HDAC expression and repressive chromatin regulators to establish an open-chromatin structure giving full access to transcriptional machineries. The key pluripotency factor Nanog cooperates with Stat3 to maintain ESC pluripotency [173]. Interestingly, HDAC inhibitors but not *NANOG* over-expression rescues complete reprogramming in the presence of STAT3 inhibition.

Finally, DNA demethylation is regulated in mammalian cells by Tet proteins (tet methylcytosine dioxygenase), which convert 5-methylcytosine (5mC) to 5-hydroxymethylcytosine (5hmC). Tet1 suppresses ESC differentiation and Tet1 knockdown leads to defects in ESC self-renewal. Tet1 up-regulation is positively regulated by Stat3 during the late-reprogramming stage [243-246].

8. Interactions between NFκB and STAT3 signaling

8.1. Synergistic NFκB and STAT3 signaling

The NFκB and STAT3 pathways are closely interconnected in regulating immune responses [247, 248]. STAT3 activation itself induces further STAT3 phosphorylation. Un-phosphorylated STAT3 that accumulates in the cell can bind to un-phosphorylated NFκB in competition with IκB. The resulting STAT3/NFκB dimer localizes to the nucleus to induce NFκB-dependent gene expression [249]. STAT3 associates with the p300/CBP (CREB-binding protein) co-activator enabling its histone acetyltranferase activity to open chromatin structures, which allows chromatin-modifying proteins to bind the DNA and activate gene transcription. [250, 251] Tyrosine-phosphorylated and acetylated STAT3 additionally binds to the NFκB precursor protein p100 and induces its processing to p52 by activation of IKKα. STAT3 then binds to the DNA-binding p52 complex to assist in the activation of target genes [252]. Both, the NFκB and STAT3 pathway synergize during terminal B-cell differentiation [253]. Phosho-p65/STAT3 dimers and phospho-STAT3/NFκB dimer complexes can bind to κB motifs. Also, phospho-STAT3 and phosho-p50 interact with each other. Soluble CD40L rapidly activates NFκB p65 and up-regulates IL10 receptors on the cell surface. This renders STAT3 more susceptible to IL-10 induced phosphorylation [161]. Macrophage activation is regulated by Toll-like receptors, JAK/STAT signaling and immunoreceptors that signal via ITAM motifs [254, 255]. These pathways have low activity levels under homeostatic conditions but are strongly activated during innate immune responses. ITAM-coupled receptors cooperate with TLRs in driving

NFκB signaling and inflammation during infections, whereas extensive ITAM activation inhibits JAK/STAT signaling to limit the immune reaction [256, 257]. Pleiotropic cytokines like interferons and IL-6 regulate the balance of pro-and anti-inflammatory functions by activating variable levels of STAT1 and STAT3 [258].

8.2. NFκB and STAT3 synergies in stem cells

NFκB and STAT3 are also part of an important stem cell pathway axis [259, 260]. A functional link between NANOG, NFκB and LIF/STAT3 signaling was shown in the maintenance of pluripotency [228]. Non-canonical NFκB signaling is activated by STAT3 through activation of IKKα and p100 processing [58]. Conversely, STAT3 inhibits TLR-induced canonical NFκB activity probably through up-regulated SOCS3. C-terminal binding of NANOG inhibits the pro-differentiation activities of canonical NFκB signaling and directly cooperates with STAT3 to maintain ESC pluripotency. NANOG and STAT3 bind to each other and synergistically activate STAT3-dependent promoters [106, 261].

The STAT3 pathway also interacts with many signaling pathways that are critically involved in the reprogramming process. For example, STAT3 signaling activates the MYC transcriptome and signals in loop with LIN28 [229]. LIN28 is expressed in undifferentiated hESC and is able to enhance the reprogramming efficiency of fibroblasts. It is down-regulated upon ESC differentiation [262-265]. Proto-oncogene tyrosine-protein kinase Src activation triggers an inflammatory response mediated by NFκB that directly activates *IL6* and *Lin28B* expression through a binding site in the first intron. IL6-mediated activation of STAT3 transcription is necessary for monocyte activation and tumorigenesis. IL6 itself further activates NFκB, thereby completing a positive NFκB-STAT3-IL6 feedback loop that links inflammation to cell transformation [229]. Constitutive STAT3 signaling maintains constitutive NFκB activity in tumors by inhibiting its nuclear export through p65 acetylation, although STAT3 signaling inhibits NFκB activation during normal immune responses [52].

9. The role of epigenetic regulators during the induction of pluripotency

9.1. The NuRD complex

A panoply of chromatin remodelers play active, regulatory roles during the reprogramming process [266, 267]. For example, the Mbd3/NuRD complex is an important epigenetic regulator that restricts the expression of key pluripotency genes [268]. MBD3 (Methyl-CpG-binding domain protein 3) is part of the NuRD (nucleosome remodeling and deacetylation) repressor complex, which mediates chromatin remodeling through histone deacetylation via HDAC1/2 and ATPase activities [269-271]. The NuRD complex interacts with methylated DNA to mediate heterochromatin formation and transcriptional silencing of ESC-specific genes. Whereas MBD2 recruits NuRD to methylated DNA, MBD3 fails to bind methylated DNA as it evolved from a methyl-CpG-binding domain to a protein–protein interaction module [272]. Mbd3 antagonizes the establishment of pluripotency and facilitates differentiation [273].

9.2. MBD3 suppression is a rate-limiting step in factor-mediated reprogramming

Recent evidence suggested that efficient reprogramming may require NuRD complex down-regulation [274]. The reprogramming factors OCT4, SOX2, KLF4 and MYC bind to MBD3, a critical component of the NURD complex. In the absence of MBD3, *SOKM* over-expression induces pluripotency with almost 100% efficiency [275]. Such reprogramming occurs within seven days in mouse cells. Once pluripotency is established, MBD3 does not appear to compromise its maintenance. The MBD3/NuRD repressor complex is probably the predominant molecular block that prevents the induction of ground-state pluripotency. Several reprogramming factors directly interact with the MBD3/NuRD complex to form a potent negative regulatory complex that restrains pluripotency gene reactivation. Thus, chromatin de-repression is of critical importance for the conversion of somatic cells into iPSC.

9.3. Bivalent histone modifications

Embryonic stem cells are not only able to maintain their undifferentiated state indefinitely, but also need to retain their ability to differentiate into various cell types [276]. The co-existence of these two features requires the combined action of signal transduction pathways, transcription factor networks, and epigenetic regulators [277]. Pluripotent gene expression has to be maintained in a way that it can be rapidly silenced upon receiving differentiation signals. The NuRD complex maintains this ESC flexibility by inducing variability in pluripotency factor expression that results in a low-expressing subpopulation of ESCs primed for differentiation [268, 278]. The control of gene expression by juxtaposition of antagonistic chromatin regulators is a common regulatory strategy in ESC, called bivalent histone modification [279, 280]. Individual promoters exhibit trimethylation of two different residues of histone H3: lysine 4 (H3K4me3) and lysine 27 (H3K27me3) [281, 282]. H3K27me3 is a repressive histone modification, whereas H3K4me3 is an activation-associated mark [283]. Both epigenetic markers have opposing effects and allow quick adjustments between ESC self-renewal and differentiation. Bivalent genes are generally transcriptionally silent in ESCs but are prone for rapid activation. MBD3 binding is enriched at bivalent genes characterized by 5hmC modifications. STAT3 binds to bivalent histone modifications and is able to switch between cellular pluripotency and differentiation [236, 284, 285].

9.4. MBD3 may prevent completion of the reprogramming process

MBD3 plays key roles in the biology of 5-hydroxy-methylcytosine (5hmC) [286]. 5hMC is an oxidation product of 5-methylcytosine (5mC) [287, 288]. MBD3 silences pluripotency genes like *Oct4* and *Nanog* through 5-hydroxy-methylation of their promoters. MBD3 binds to 5hmC in cooperation with Tet1 to regulate 5hmC-marked genes, but does not interact with 5mC. Mbd3 interaction with 5hmC recruits NuRD to its targets resulting in gene repression. Knockdown of the MBD3/NuRD complex affects the expression of 5hmC-marked genes [289]. Mbd3 acts upstream of Nanog and may block the transition from partially to fully reprogrammed iPSC by silencing Nanog. *Nanog* overexpression was dominant over Mbd3 knockdown in the induction of efficient reprogramming and is in general sufficient to maintain mESC pluripotency. Mbd3 depletion facilitates the transcription of *Oct4* and *Nanog* and leads to the generation of iPSC and chimeric mice even in the absence of Sox2 or c-Myc [290]. The depletion of Mbd3/NuRD does not replace Oct4 during iPSC formation as reprogramming did not occur with Klf4 and c-Myc alone. Mbd3-dependent silencing of pluripotency factors occurs during ESC differentiation. This involves NuRD-dependent deacetylation of H3K27 required for the binding of the polycomb repressive complex two. NuRD-dependent silencing of pluripotency genes prevents the de-differentiation of somatic cells. In the absence of Mbd3, NuRD disassembles, which lowers this epigenetic barrier and allows the activation of pluripotency genes. Drug-induced down-regulation of Mbd3/NuRD may greatly improve the efficiency and fidelity of reprogramming [291].

9.5. STAT3-MBD3 counteractions

Stat3 promotes the expression of self-renewal transcription factors and opposes NURD-mediated repression of several hundred target genes in ESCs. The opposing functions of Stat3 and NuRD maintain variability in the levels of key self-renewal transcription factors. Stat3, but not NuRD, is the rate-limiting factor for pluripotency gene expression. Self-renewing ESC face a barrier that prohibits differentiation. NuRD constrains this barrier within a range that can be overcome when self-renewal signals are withdrawn [268, 278, 292]. Mbd3/NuRD-mediated gene silencing is a critical determinant of lineage commitment in embryonic stem cells and allows cells to exhibit pluripotency and self-renewal. Mbd3-deficient ESC show

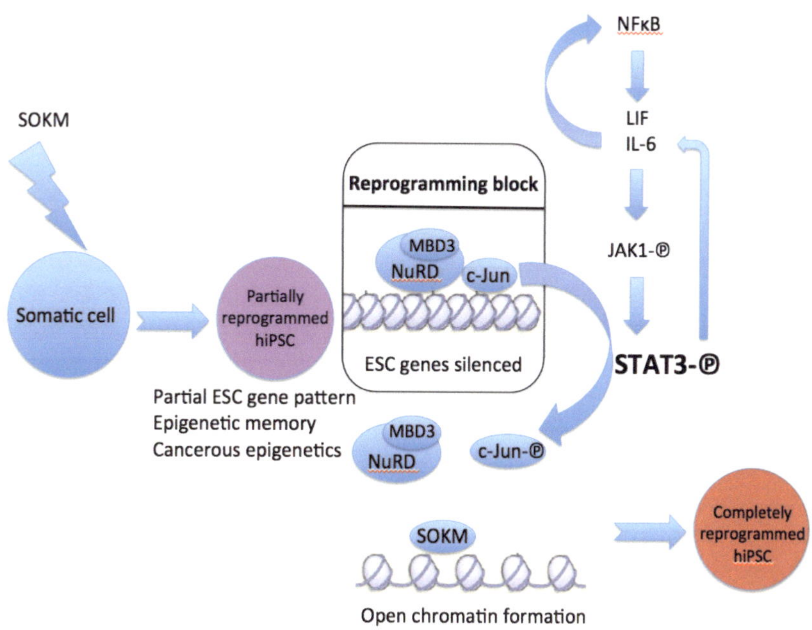

Figure 2. The master reprogramming factor STAT3 may overcome an unknown reprogramming block by inducing an open chromatin formation that facilitates the pluripotency factors SOKM to bind to ESC gene promoters. We hypothesize that upstream inflammatory signals mediated by NFκB signaling may facilitate STAT3 to de-repress the NuRD complex via c-Jun.

persistent self-renewal even in the absence of Lif. They are able to undergo the initial steps of differentiation, but their ability for lineage commitment is severely compromised. They fail to downregulate undifferentiated cell markers as well as upregulate differentiation markers [293]. Stat3 has many downstream effectors like the proto-oncogene c-Jun that is part of the AP-1 complex [194]. The transactivation domain of un-phosphorylated c-Jun recruits Mbd3/NuRD to AP-1 target genes to mediate gene repression. This repression is relieved by c-Jun N-terminal phosphorylation or Mbd3 depletion. Upon JNK activation, NuRD dissociates from c-Jun, which results in de-repression of target gene transcription. Termination of the JNK signal induces Mbd3/NuRD re-binding to un-phosphorylated c-Jun and cessation of target gene expression (Figure 2) [199].

10. Conclusions

In this review, we have discussed a potentially novel link between inflammatory pathways and efficient cell reprogramming. In this context, our group reported that bone marrow stromal-primed human myeloid cell progenitors are significantly more receptive to reprogramming stimuli than other cell types [20]. Myeloid cells harbor a unique epigenetic plasticity that allows them to quickly respond to a plethora of pathogens. They are innately equipped to transcriptionally and epigenetically activate key inflammatory pathways via an interconnected NFκB and STAT3 signaling machinery [294]. Both pathways act as epigenetic modifiers during normal inflammation stimulation, and both are also known to promote ESC pluripotency by inducing an open chromatin state that allows other transcription factors to regulate cell fates [236]. This epigenetic remodeling may prove crucial for efficient reprogramming, as well as the generation of high quality iPSC that resemble ESC without excessive epigenetic memory of their cell of origin [295].

Moreover, Stat3 is a master reprogramming factor that is able to dominantly instruct pluripotency, yet is also inherently interconnected with inflammatory signaling cascades (Figure 2). It binds to bivalent histone modifications, and allows rapid transitions between pluripotency and differentiation [193]. The NFκB pathway acts in synergy with downstream STAT3 signaling, whereby non-canonical NFκB signaling maintains pluripotency through epigenetic silencing of differentiation genes and canonical NFκB signaling promotes cell differentiation [296]. Finally, recent evidence suggests that strong chromatin repression by the NuRD complex is a key rate-limiting factor during reprogramming to pluripotency. This important complex may normally function to ensure that differentiated cells do not reactivate pluripotency genes, which might enable tumorigenesis [268]. We propose the hypothesis that NuRD complex silencing might be more easily achieved through the activation of inflammatory pathways in receptive cells such as those from the myeloid lineage.

It remains to be elucidated how all these processes are inter-regulated. It will be especially important to link reprogramming efficiency with the resulting quality of the pluripotent state achieved in hiPSC. We hypothesize that epigenetic plasticity in inflammatory cells that normally allows chromatin accessibility to the transcriptional machinery, could be manipulated to facilitate a complete erasure of the donor epigenetic memory during factor-mediated reprogramming. Additionally, preventing cancerous epigenetic patterns in iPSC via more accurate high-fidelity reprogramming methods will be the foundation for future clinical applications [13]. Finally, the basic understanding of pluripotency induction may also give us a better understanding of how tumor-initiating cells arise and how they can be eradicated to prevent tumor relapse, thus potentially opening a new era of cancer treatments.

Acknowledgements

JRA was supported by a fellowship from the German Research Foundation (DFG, DJ 71/1-1). ETZ was supported by grants from the NIH/NHLBI U01HL099775 and the Maryland Stem Cell Research Fund (2011-MSCRF-II-0008-00; 2007-MSCRF-II-0379-00). We would like to thank Dr. Alan Friedman for assistance in reading and editing the manuscript.

Author details

Jasmin Roya Agarwal and Elias T. Zambidis

Institute for Cell Engineering and Sidney Kimmel Comprehensive Cancer Center, The Johns Hopkins University School of Medicine, Baltimore, USA

References

[1] Chin MH, Mason MJ, Xie W, Volinia S, Singer M, Peterson C, et al. Induced pluripotent stem cells and embryonic stem cells are distinguished by gene expression signatures. Cell Stem Cell. 2009;5(1):111–23.

[2] Choi K-D, Yu J, Smuga-Otto K, Salvagiotto G, Rehrauer W, Vodyanik M, et al. Hematopoietic and endothelial differentiation of human induced pluripotent stem cells. Stem Cells. 2009;27(3):559–67.

[3] Feng Q, Lu S-J, Klimanskaya I, Gomes I, Kim D, Chung Y, et al. Hemangioblastic derivatives from human induced pluripotent stem cells exhibit limited expansion and early senescence. Stem Cells. 2010;28(4):704–12.

[4] Burridge PW, Thompson S, Millrod MA, Weinberg S, Yuan X, Peters A, et al. A universal system for highly efficient cardiac differentiation of human induced pluripotent stem cells that eliminates interline variability. PloS One. 2011;6(4):e18293.

[5] Park TS, Zimmerlin L, Zambidis ET. Efficient and simultaneous generation of hematopoietic and vascular progenitors from human induced pluripotent stem cells. Cytometry A. 2012;38(1):114-26.

[6] Park TS, Bhutto I, Zimmerlin L, Huo JS, Nagaria P, Miller D, et al. Vascular Progenitors from Cord Blood-Derived iPSC Possess Augmented Capacity for Regenerating Ischemic Retinal Vasculature. Circulation. 2013 Oct 25; Epub ahead of print.

[7] Takahashi K, Yamanaka S. Induction of Pluripotent Stem Cells from Mouse Embryonic and Adult Fibroblast Cultures by Defined Factors. Cell. 2006;126(4):663–76.

[8] Simara P, Motl JA, Kaufman DS. Pluripotent stem cells and gene therapy. Translational research. 2013;161(4):284–92.

[9] Oh Y, Wei H, Ma D, Sun X, Liew R. Clinical applications of patient-specific induced pluripotent stem cells in cardiovascular medicine. Heart. 2012;98(6):443–9.

[10] Kaufman DS. Toward clinical therapies using hematopoietic cells derived from human pluripotent stem cells. Blood. 2009;114(17):3513–23.

[11] Lister R, Pelizzola M, Kida YS, Hawkins RD, Nery JR, Hon G, et al. Hotspots of aberrant epigenomic reprogramming in human induced pluripotent stem cells. Nature. 2011;471(7336):68–73.

[12] Carey BW, Markoulaki S, Hanna JH, Faddah DA, Buganim Y, Kim J, et al. Reprogramming factor stoichiometry influences the epigenetic state and biological properties of induced pluripotent stem cells. Cell Stem Cell. 2011;9(6):588–98.

[13] Easwaran H, Johnstone SE, Van Neste L, Ohm J, Mosbruger T, Wang Q, et al. A DNA hypermethylation module for the stem/progenitor cell signature of cancer. Genome Research. 2012 May 1;22(5):837–49.

[14] Ruiz S, Diep D, Gore A, Panopoulos AD, Montserrat N, Plongthongkum N, et al. Identification of a specific reprogramming-associated epigenetic signature in human induced pluripotent stem cells. Proceedings of the National Academy of Sciences. 2012;109(40):16196–201.

[15] Kim K, Doi A, Wen B, Ng K, Zhao R, Cahan P, et al. Epigenetic memory in induced pluripotent stem cells. Nature. 2010;467(7313):285–90.

[16] Kim K, Zhao R, Doi A, Ng K, Unternaehrer J, Cahan P, et al. Donor cell type can influence the epigenome and differentiation potential of human induced pluripotent stem cells. Nature Biotechnology. 2011;29(12):1117–9.

[17] Bar-Nur O, Russ HA, Efrat S, Benvenisty N. Epigenetic memory and preferential lineage-specific differentiation in induced pluripotent stem cells derived from human pancreatic islet beta cells. Cell Stem Cell. 2011;9(1):17–23.

[18] Hu K, Yu J, Suknuntha K, Tian S, Montgomery K, Choi KD, et al. Efficient generation of transgene-free induced pluripotent stem cells from normal and neoplastic bone marrow and cord blood mononuclear cells. Blood. 2011 Apr 7;117(14):e109–19.

[19] Yu J, Chau KF, Vodyanik MA, Jiang J, Jiang Y. Efficient feeder-free episomal reprogramming with small molecules. PloS One. 2011;6(3):e17557.

[20] Park TS, Huo JS, Peters A, Talbot CC, Verma K, Zimmerlin L, et al. Growth factor-activated stem cell circuits and stromal signals cooperatively accelerate non-integrated iPSC reprogramming of human myeloid progenitors. PloS One. 2012;7(8):e42838.

[21] Malik N, Rao MS. A review of the methods for human iPSC derivation. Methods Molecular Biology. 2013;997:23–33.

[22] Okita K, Matsumura Y, Sato Y, Okada A, Morizane A, Okamoto S, et al. A more efficient method to generate integration-free human iPS cells. Nature Methods. 2011;8(5):409–12.

[23] Nakagawa M, Koyanagi M, Tanabe K, Takahashi K, Ichisaka T, Aoi T, et al. Generation of induced pluripotent stem cells without Myc from mouse and human fibroblasts. Nature Biotechnology. 2008;26(1):101–6.

[24] Yu J, Hu K, Smuga-Otto K, Tian S, Stewart R, Slukvin II, et al. Human induced pluripotent stem cells free of vector and transgene sequences. Science. 2009;324(5928):797–801.

[25] Huangfu D, Maehr RE, Guo W, Eijkelenboom A, Snitow M, Chen AE, et al. Induction of pluripotent stem cells by defined factors is greatly improved by small-molecule compounds. Nature Biotechnology. 2008;26(7):795–7.

[26] Kleger A, Mahaddalkar PU, Katz S-F, Lechel AE, Joo JY, Loya K, et al. Increased reprogramming capacity of mouse liver progenitor cells, compared with differentiated liver cells, requires the BAF complex. Gastroenterology. 2012;142(4):907–17.

[27] Niu W, Zang T, Zou Y, Fang S, Smith DK, Bachoo R, et al. In vivo reprogramming of astrocytes to neuroblasts in the adult brain. Nature Cell Biology. 2013;15(10):1164–75.

[28] Grande A, Sumiyoshi K, L o pez-Ju a rez A, Howard J, Sakthivel B, Aronow B, et al. Environmental impact on direct neuronal reprogramming in vivo in the adult brain. Nature Communications. 2013;4:2373.

[29] Zou X-Y, Yang H-Y, Yu Z, Tan X-B, Yan X, Huang GT-J. Establishment of transgene-free induced pluripotent stem cells reprogrammed from human stem cells of apical papilla for neural differentiation. Stem Cell Research & Therapy. 2012;3(5):43.

[30] Chen J, Lin M, Foxe JJ, Pedrosa E, Hrabovsky A, Carroll R, et al. Transcriptome comparison of human neurons generated using induced pluripotent stem cells derived from dental pulp and skin fibroblasts. PloS One. 2013;8(10):e75682.

[31] Staerk J, Dawlaty MM, Gao Q, Maetzel D, Hanna J, Sommer CA, et al. Reprogramming of human peripheral blood cells to induced pluripotent stem cells. Cell Stem Cell. 2010;7(1):20–4.

[32] Loh Y-H, Agarwal S, Park I-H, Urbach A, Huo H, Heffner GC, et al. Generation of induced pluripotent stem cells from human blood. Blood. 2009;113(22):5476–9.

[33] Haase A, Olmer R, Schwanke K, Wunderlich S, Merkert S, Hess C, et al. Generation of induced pluripotent stem cells from human cord blood. Cell Stem Cell. 2009;5(4): 434–41.

[34] Springer TA. Adhesion receptors of the immune system. Nature. 1990;346(6283):425–34.

[35] Medzhitov R, Janeway C. Innate Immunity. New England Journal Medicine. 2000;343(5):338-44.

[36] Alberts B, Johnson A, Lewis J, Raff M, Roberts K, Walter P. Molecular Biology of the Cell, 5th edition. Taylor&Francis; 2007.

[37] Lanzavecchia A, Bernasconi N, Traggiai E, Ruprecht CR, Corti D, Sallusto F. Understanding and making use of human memory B cells. Immunological Reviews. 2006;211:303–9.

[38] Eminli S, Foudi A, Stadtfeld M, Maherali N, Ahfeldt T, Mostoslavski G, Hock H, Hochedlinger K. differentiation stage determines potential of hematopoietic cells for reprogramming into induced pluripotent stem cells. Nature Genetics. 2009;41(9): 970-976.

[39] Gaspar-Maia A, Alajem A, Meshorer E, Ramalho-Santos M. Open chromatin in pluripotency and reprogramming. Nature reviews Molecular Cell Biology. 2011;12(1):36–47.

[40] Lee J, Sayed N, Hunter A, Au KF, Wong WH, Mocarski ES, et al. Activation of innate immunity is required for efficient nuclear reprogramming. Cell. 2012;151(3):547–58.

[41] Kawamoto H, Wada H, Katsura Y. A revised scheme for developmental pathways of hematopoietic cells: the myeloid-based model. International Immunology. 2010;22(2): 65–70.

[42] Zingarelli B. Nuclear factor-kappaB. Critical Care Medicine. 2005;33(12):414–6.

[43] Abraham E. NF-kappaB activation. Critical Care Medicine. 2000;28(4):100–4.

[44] Salminen A, Huuskonen J, Ojala J, Kauppinen A, Kaarniranta K, Suuronen T. Activation of innate immunity system during aging: NF-kB signaling is the molecular culprit of inflamm-aging. Ageing Research Reviews. 2008;7(2):83–105.

[45] Luo J-L, Kamata H, Karin M. IKK/NF-kappaB signaling: balancing life and death--a new approach to cancer therapy. The Journal of Clinical Investigation. 2005;115(10): 2625–32.

[46] Vanden Berghe W, Ndlovu MN, Hoya-Arias R, Dijsselbloem N, Gerlo S, Haegeman G. Keeping up NF-kappaB appearances: epigenetic control of immunity or inflammation-triggered epigenetics. Biochemical Pharmacology. 2006;72(9):1114–31.

[47] Gilmore TD. Introduction to NF-kappaB: players, pathways, perspectives. Oncogene. 2006;25(51):6680–4.

[48] Tak PP, Firestein GS. NF-kappaB: a key role in inflammatory diseases. The Journal of Clinical Investigation. 2001;107(1):7–11.

[49] Chen ZJ. Ubiquitin signalling in the NF-kappaB pathway. Nature Cell Biology. 2005;7(8):758–65.

[50] Kawai T, Akira S. Signaling to NF-kappaB by Toll-like receptors. Trends in Molecular Medicine. 2007;13(11):460–9.

[51] Chen Lf, Fischle W, Verdin E, Greene WC. Duration of nuclear NF-kappaB action regulated by reversible acetylation. Science. 2001;293(5535):1653–7.

[52] Lee H, Herrmann A, Deng J-H, Kujawski M, Niu G, Li Z, et al. Persistently activated Stat3 maintains constitutive NF-kappaB activity in tumors. Cancer Cell. 2009;15(4): 283–93.

[53] Dyson HJ, Komives EA. Role of disorder in IkappaB-NFkappaB interaction. IUBMB Life. 2012;64(6):499–505.

[54] Panwalkar A, Verstovsek S, Giles F. Nuclear factor-kappaB modulation as a therapeutic approach in hematologic malignancies. Cancer. 2004;100(8):1578–89.

[55] Imani Fooladi AA, Mousavi SF, Seghatoleslami S, Yazdani S, Nourani MR. Toll-like receptors: role of inflammation and commensal bacteria. Inflammation & Allergy Drug Targets. 2011;10(3):198–207.

[56] Li X, Jiang S, Tapping RI. Toll-like receptor signaling in cell proliferation and survival. Cytokine. 2010;49(1):1–9.

[57] Carmody RIJ, Chen YH. Nuclear factor-kappaB: activation and regulation during toll-like receptor signaling. Cellular & Molecular Immunology. 2007;4(1):31–41.

[58] Sun S-C. Non-canonical NF-kappaB signaling pathway. Cell Research. 2011;21(1):71–85.

[59] Endo T, Nishio M, Enzler T, Cottam HB, Fukuda T, James DF, et al. BAFF and APRIL support chronic lymphocytic leukemia B-cell survival through activation of the canonical NF-kappaB pathway. Blood. 2007;109(2):703–10.

[60] Meichle, Jurkat. Protein kinase C-independent activation of nuclear factor kappa B by tumor necrosis factor. J Biol Chem. 1990;265(14):8339–43.

[61] Mordm u ller B, Krappmann D, Esen M, Wegener E, Scheidereit C. Lymphotoxin and lipopolysaccharide induce NF-kappaB-p52 generation by a co-translational mechanism. EMBO reports. 2003;4(1):82–7.

[62] Moschonas A, Ioannou M, Eliopoulos AG. CD40 stimulates a "feed-forward" NF-kappaB-driven molecular pathway that regulates IFN-beta expression in carcinoma cells. Journal of Immunology. 2012;188(11):5521–7.

[63] H a cker H, Tseng P-H, Karin M. Expanding TRAF function: TRAF3 as a tri-faced immune regulator. Nature Reviews Immunology. 2011;11(7):457–68.

[64] Sinha SK, Zachariah S, Qui n ones HI, Shindo M, Chaudhary PM. Role of TRAF3 and-6 in the activation of the NF-kappa B and JNK pathways by X-linked ectodermal dysplasia receptor. The Journal of Biological Chemistry. 2002;277(47):44953–61.

[65] Morrison MD, Reiley W, Zhang M, Sun SC. An atypical TRAF-binding motif of BAFF receptor mediates induction of the noncanonical NF-kB signaling pathway. Journal of Biological Chemistry. 2005.

[66] Lu J, Hou R, Booth CJ, Yang S-H, Snyder M. Defined culture conditions of human embryonic stem cells. Proceedings of the National Academy of Sciences. 2006;103(15):5688–93.

[67] Ni C-Z, Oganesyan G, Welsh K, Zhu X, Reed JC, Satterthwait AC, et al. Key molecular contacts promote recognition of the BAFF receptor by TNF receptor-associated factor 3: implications for intracellular signaling regulation. Journal of Immunology. 2004;173(12):7394–400.

[68] Hu H, Brittain GC, Chang J-H, Puebla-Osorio N, Jin J, Zal A, et al. OTUD7B controls non-canonical NF-kappaB activation through deubiquitination of TRAF3. Nature. 2013;494(7437):371-4.

[69] Kehry MR. CD40-mediated signaling in B cells. Balancing cell survival, growth, and death. Journal of Immunology. 1996;156(7):2345–8.

[70] Gelbmann CM, Leeb SN, Vogl D, Maendel M, Herfarth H, Sch o lmerich J, et al. Inducible CD40 expression mediates NFkappaB activation and cytokine secretion in human colonic fibroblasts. Gut. 2003;52(10):1448–56.

[71] Tsukamoto N, Kobayashi N, Azuma S, Yamamoto T, Inoue J. Two differently regulated nuclear factor B activation pathways triggered by the cytoplasmic tail of CD40. Proceedings of the National Academy of Sciences. 1999;96(4):1234–9.

[72] Zarnegar B, He JQ, Oganesyan G, Hoffmann A, Baltimore D, Cheng G. Unique CD40-mediated biological program in B cell activation requires both type 1 and type

2 NF-kappaB activation pathways. Proceedings of the National Academy of Sciences. 2004;101(21):8108–13.

[73] Zhao C, Xiu Y, Ashton J, Xing L, Morita Y, Jordan CT, et al. Noncanonical NF-kappaB signaling regulates hematopoietic stem cell self-renewal and microenvironment interactions. Stem Cells. 2012;30(4):709–18.

[74] Jost PJ, Ruland JUR. Aberrant NF-kappaB signaling in lymphoma: mechanisms, consequences, and therapeutic implications. Blood. 2007;109(7):2700–7.

[75] Senftleben U, Karin M. The IKK/NF-kappa B pathway. Critical Care Medicine. 2002;30(1 Suppl):S18–26.

[76] Chauhan D, Uchiyama H, Akbarali Y, Urashima M, Yamamoto K, Libermann TA, et al. Multiple myeloma cell adhesion-induced interleukin-6 expression in bone marrow stromal cells involves activation of NF-kappa B. Blood. 1996;87(3):1104–12.

[77] Takeda K, Harada Y, Watanabe R, Inutake Y, Ogawa S, Onuki K, et al. CD28 stimulation triggers NF-kappaB activation through the CARMA1-PKCtheta-Grb2/Gads axis. International Immunology. 2008;20(12):1507–15.

[78] Coope HJ, Atkinson PGP, Huhse B, Belich M, Janzen J, Holman MJ, et al. CD40 regulates the processing of NF-kappaB2 p100 to p52. The EMBO Journal. 2002;21(20): 5375–85.

[79] Riha P, Rudd CE. CD28 co-signaling in the adaptive immune response. Self Nonself. 2010;1(3):231–40.

[80] Muscolini, Jurkat. A novel association between filamin A and NF-kappaB inducing kinase couples CD28 to inhibitor of NF-kappaB kinase alpha and NF-kappaB activation. Immunology Letters. 2011;136(2):203–12.

[81] Piccolella, CD28. Vav-1 and the IKK alpha subunit of I kappa B kinase functionally associate to induce NF-kappa B activation in response to CD28 engagement. Journal of Immunology. 2003;170(6):2895–903.

[82] Wang Z, Oron E, Nelson B, Razis S, Ivanova N. Distinct lineage specification roles for NANOG, OCT4, and SOX2 in human embryonic stem cells. Cell Stem Cell. 2012;10(4):440–54.

[83] Chan KK-K, Zhang J, Chia N-Y, Chan Y-S, Sim HS, Tan KS, et al. KLF4 and PBX1 directly regulate NANOG expression in human embryonic stem cells. Stem Cells. 2009;27(9):2114–25.

[84] Zeng X. Human embryonic stem cells: mechanisms to escape replicative senescence? Stem Cell Reviews. 2007;3(4):270–9.

[85] Gourronc FA, Klingelhutz AJ. Therapeutic opportunities: telomere maintenance in inducible pluripotent stem cells. Mutation Research. 2012;730(1-2):98–105.

[86] Klimanskaya I, Chung Y, Meisner L, Johnson J, West MD, Lanza R. Human embryonic stem cells derived without feeder cells. Lancet. 365(9471):1636–41.

[87] Dutta D, Ray S, Home P, Larson M, Wolfe MW, Paul S. Self-renewal versus lineage commitment of embryonic stem cells: protein kinase C signaling shifts the balance. Stem Cells. 2011;29(4):618–28.

[88] Molinero, CD28. High TCR stimuli prevent induced regulatory T cell differentiation in a NF-kappaB-dependent manner. Journal of Immunology. 2011;186(8):4609–17.

[89] Yang C, Atkinson SP, Vilella F, Lloret M, Armstrong L, Mann DA, et al. Opposing putative roles for canonical and noncanonical NFkappaB signaling on the survival, proliferation, and differentiation potential of human embryonic stem cells. Stem Cells. 2010;28(11):1970–80.

[90] Kang H-B, Kim Y-E, Kwon H-J, Sok D-E, Lee Y. Enhancement of NF-kappaB expression and activity upon differentiation of human embryonic stem cell line SNUhES3. Stem Cells and Development. 2007;16(4):615–23.

[91] De Molfetta GA, Luciola Zanette D, Alexandre Panepucci R, Dos Santos ARD, da Silva WA, Antonio Zago M. Role of NFKB2 on the early myeloid differentiation of CD34+hematopoietic stem/progenitor cells. Differentiation. 2010;80(4-5):195–203.

[92] Wang D, Paz-Priel I, Friedman AD. NF-kappa B p50 regulates C/EBP alpha expression and inflammatory cytokine-induced neutrophil production. Journal of Immunology. 2009;182(9):5757–62.

[93] Nakata S, Matsumura I, Tanaka H, Ezoe S, Satoh Y, Ishikawa J, et al. NF-kappaB family proteins participate in multiple steps of hematopoiesis through elimination of reactive oxygen species. The Journal of Biological Chemistry. 2004;279(53):55578–86.

[94] Bottero V, Withoff S, Verma IM. NF-kappaB and the regulation of hematopoiesis. Cell Death and Differentiation. 2006;13(5):785–97.

[95] Schwarzer R, Jundt F. Notch and NF-kappaB signaling pathways in the biology of classical Hodgkin lymphoma. Current Molecular Medicine. 2011;11(3):236–45.

[96] Maniati E, Bossard M, Cook N, Candido JB, Emami-Shahri N, Nedospasov SA, et al. Crosstalk between the canonical NF-kappaB and Notch signaling pathways inhibits Ppargamma expression and promotes pancreatic cancer progression in mice. The Journal of Clinical Investigation. 2011;121(12):4685–99.

[97] Cheng P, Zlobin A, Volgina V, Gottipati S, Osborne B, Simel EJ, et al. Notch-1 regulates NF-kappaB activity in hemopoietic progenitor cells. Journal of Immunology. 2001;167(8):4458–67.

[98] Wang J, Shelly L, Miele L, Boykins R, Norcross MA, Guan E. Human Notch-1 inhibits NF-kappa B activity in the nucleus through a direct interaction involving a novel domain. Journal of Immunology. 2001;167(1):289–95.

[99] Lin G, Xu R-H. Progresses and challenges in optimization of human pluripotent stem cell culture. Current Stem Cell Research & therapy. 2010;5(3):207–14.

[100] Darr H, Benvenisty N. Human embryonic stem cells: the battle between self-renewal and differentiation. Regenerative Medicine. 2006;1(3):317–25.

[101] Dvorak P, Dvorakova D, Hampl A. Fibroblast growth factor signaling in embryonic and cancer stem cells. FEBS letters. 2006;580(12):2869–74.

[102] Byrd VM, Ballard DW, Miller GG, Thomas JW. Fibroblast growth factor-1 (FGF-1) enhances IL-2 production and nuclear translocation of NF-kappaB in FGF receptor-bearing Jurkat T cells. Journal of Immunology. 1999;162(10):5853–9.

[103] Stein SJ, Baldwin AS. Deletion of the NF-kappaB subunit p65/RelA in the hematopoietic compartment leads to defects in hematopoietic stem cell function. Blood. 2013.

[104] Panepucci RA, Calado RT, Rocha V, Proto-Siqueira R, Silva WA, Zago MA. Higher expression of transcription targets and components of the nuclear factor-kappaB pathway is a distinctive feature of umbilical cord blood CD34+precursors. Stem Cells. 2007;25(1):189–96.

[105] Luningschror P, St o cker B, Kaltschmidt B, Kaltschmidt C. miR-290 cluster modulates pluripotency by repressing canonical NF-kappaB signaling. Stem Cells. 2012;30(4):655–64.

[106] Torres J, Watt FM. Nanog maintains pluripotency of mouse embryonic stem cells by inhibiting NFkappaB and cooperating with Stat3. Nature Cell Biology. 2008;10(2): 194–201.

[107] Taylor T, Kim Y-J, Ou X, Derbigny W, Broxmeyer HE. Toll-like receptor 2 mediates proliferation, survival, NF-kappaB translocation, and cytokine mRNA expression in LIF-maintained mouse embryonic stem cells. Stem Cells and Development. 2010;19(9):1333–41.

[108] Chiffoleau E, Kobayashi T, Walsh MC, King CG, Walsh PT, Hancock WW, et al. TNF receptor-associated factor 6 deficiency during hemopoiesis induces Th2-polarized inflammatory disease. Journal of Immunology. 2003;171(11):5751–9.

[109] Yaddanapudi K, De Miranda J, Hornig M, Lipkin WI. Toll-like receptor 3 regulates neural stem cell proliferation by modulating the Sonic Hedgehog pathway. PloS One. 2011;6(10):e26766.

[110] Chinen J, Notarangelo LD, Shearer WT. Advances in basic and clinical immunology in 2012. The Journal of Allergy and Clinical Immunology. 2013;131(3):675–82.

[111] Takase O, Yoshikawa M, Idei M, Hirahashi J, Fujita T, Takato T, et al. The Role of NF-kappaB Signaling in the Maintenance of Pluripotency of Human Induced Pluripotent Stem Cells. PloS One. 2013;8(2):e56399.

[112] Georgopoulos K. Haematopoietic cell-fate decisions, chromatin regulation and ikaros. Nature Reviews Immunology. 2002;2(3):162–74.

[113] Smale ST, Fisher AG. Chromatin structure and gene regulation in the immune system. Annual Review of Immunology. 2002;20:427–62.

[114] Galm O, Herman JG, Baylin SB. The fundamental role of epigenetics in hematopoietic malignancies. Blood Reviews. 2006;20(1):1–13.

[115] Sharma S, Kelly TK, Jones PA. Epigenetics in cancer. Carcinogenesis. 2010;31(1):27–36.

[116] Meshorer E, Misteli T. Chromatin in pluripotent embryonic stem cells and differentiation. Nature Reviews Molecular Cell Biology. 2006;7(7):540–6.

[117] Wirtz S, Becker C, Fantini MC, Nieuwenhuis EE, Tubbe I, Galle PR, et al. EBV-induced gene 3 transcription is induced by TLR signaling in primary dendritic cells via NF-kappa B activation. Journal of Immunology. 2005;174(5):2814–24.

[118] Gaudreault E, Fiola SEP, Olivier M, Gosselin J. Epstein-Barr virus induces MCP-1 secretion by human monocytes via TLR2. Journal of Virology. 2007;81(15):8016–24.

[119] Severa M, Giacomini E, Gafa V, Anastasiadou E, Rizzo F, Corazzari M, et al. EBV stimulates TLR-and autophagy-dependent pathways and impairs maturation in plasmacytoid dendritic cells: Implications for viral immune escape. European Journal of Immunology. 2013;43(1):147-58.

[120] Martin HJ, Lee JM, Walls D, Hayward SD. Manipulation of the toll-like receptor 7 signaling pathway by Epstein-Barr virus. Journal of Virology. 2007;81(18):9748–58.

[121] Shimada M, Ishimoto T, Lee PY, Lanaspa MA, Rivard CJ, Roncal-Jimenez CA, et al. Toll-like receptor 3 ligands induce CD80 expression in human podocytes via an NF-kappaB-dependent pathway. Nephrology Dialysis Transplantation. 2012;27(1):81–9.

[122] Salaun B, Coste I, Rissoan M-C, Lebecque SJ, Renno T. TLR3 can directly trigger apoptosis in human cancer cells. Journal of Immunology. 2006;176(8):4894–901.

[123] Gohda J, Matsumura T, Inoue J-I. Cutting edge: TNFR-associated factor (TRAF) 6 is essential for MyD88-dependent pathway but not toll/IL-1 receptor domain-containing adaptor-inducing IFN-beta (TRIF)-dependent pathway in TLR signaling. Journal of Immunology. 2004;173(5):2913–7.

[124] Biran A, Meshorer E. Concise review: chromatin and genome organization in reprogramming. Stem Cells. 2012;30(9):1793–9.

[125] Hawkins RD, Hon GC, Lee LK, Ngo Q, Lister R, Pelizzola M, et al. Distinct epigenomic landscapes of pluripotent and lineage-committed human cells. Cell Stem Cell. 2010;6(5):479–91.

http://dx.doi.org/10.5772/57602

[126] Org TON, Chignola F, Het e nyi C, Gaetani M, Rebane A, Liiv I, et al. The autoimmune regulator PHD finger binds to non-methylated histone H3K4 to activate gene expression. EMBO Reports. 2008;9(4):370–6.

[127] Milutinovic S, D'Alessio AC, Detich N, Szyf M. Valproate induces widespread epigenetic reprogramming which involves demethylation of specific genes. Carcinogenesis. 2007;28(3):560–71.

[128] Anokye-Danso F, Trivedi CM, Juhr D, Gupta M, Cui Z, Tian Y, et al. Highly efficient miRNA-mediated reprogramming of mouse and human somatic cells to pluripotency. Cell Stem Cell. 2011;8(4):376–88.

[129] Dekker FJ, Haisma HJ. Histone acetyl transferases as emerging drug targets. Drug Discovery Today. 2009;14(19-20):942–8.

[130] Imhof A, Yang XJ, Ogryzko VV, Nakatani Y, Wolffe AP, Ge H. Acetylation of general transcription factors by histone acetyltransferases. Current Biology. 1997;7(9):689–92.

[131] Perkins ND, Felzien LK, Betts JC, Leung K, Beach DH, Nabel GJ. Regulation of NF-kappaB by cyclin-dependent kinases associated with the p300 coactivator. Science. 1997;275(5299):523–7.

[132] Sheppard KA, Rose DW, Haque ZK, Kurokawa R, McInerney E, Westin S, et al. Transcriptional activation by NF-kappaB requires multiple coactivators. Molecular and Cellular Biology. 1999;19(9):6367–78.

[133] Chen X, Gazzar El M, Yoza BK, McCall CE. The NF-kappaB factor RelB and histone H3 lysine methyltransferase G9a directly interact to generate epigenetic silencing in endotoxin tolerance. The Journal of Biological Chemistry. 2009 Oct 9;284(41):27857–65.

[134] Roger T, Lugrin JEROM, Le Roy D, Goy GEV, Mombelli M, Koessler T, et al. Histone deacetylase inhibitors impair innate immune responses to Toll-like receptor agonists and to infection. Blood. 2011;117(4):1205–17.

[135] Pearce EL, Shen H. Making sense of inflammation, epigenetics, and memory CD8+T-cell differentiation in the context of infection. Immunological Reviews. 2006;211:197–202.

[136] McCall CE, Yoza B, Liu T, Gazzar El M. Gene-Specific Epigenetic Regulation in Serious Infections with Systemic Inflammation. J Innate Immun. 2010;2(5):395–405.

[137] Niwa H. How is pluripotency determined and maintained? Development.; 2007 Feb; 134(4):635–46.

[138] O'Neill LAJ. "Transflammation": When Innate Immunity Meets Induced Pluripotency. Cell. 2012 Oct;151(3):471–3.

[139] Heinrich PC, Behrmann I, M u ller-Newen G, Schaper F, Graeve L. Interleukin-6-type cytokine signalling through the gp130/Jak/STAT pathway. The Biochemical Journal. 1998;334 (Pt 2:297–314.

[140] Imada K, Leonard WJ. The Jak-STAT pathway. Molecular Immunology. 37(1-2):1–11.

[141] Takemoto S, Mulloy JC, Cereseto A, Migone TS, Patel BK, Matsuoka M, et al. Proliferation of adult T cell leukemia/lymphoma cells is associated with the constitutive activation of JAK/STAT proteins. Proceedings of the National Academy of Sciences. 1997;94(25):13897–902.

[142] Shuai K, Liu B. Regulation of JAK-STAT signalling in the immune system. Nature Reviews Immunology. 2003;3(11):900–11.

[143] Sherry MM, Reeves A, Wu JK, Cochran BH. STAT3 is required for proliferation and maintenance of multipotency in glioblastoma stem cells. Stem cells. 2009;27(10):2383–92.

[144] Ram PT, Iyengar R. G protein coupled receptor signaling through the Src and Stat3 pathway: role in proliferation and transformation. Oncogene. 2001;20(13):1601–6.

[145] Hirano T, Ishihara K, Hibi M. Roles of STAT3 in mediating the cell growth, differentiation and survival signals relayed through the IL-6 family of cytokine receptors. Oncogene. 2000;19(21):2548–56.

[146] Al Zaid Siddiquee K, Turkson J. STAT3 as a target for inducing apoptosis in solid and hematological tumors. Cell Research. 2008;18(2):254–67.

[147] Schindler C, Levy DE, Decker T. JAK-STAT signaling: from interferons to cytokines. The Journal of Biological Chemistry. 2007;282(28):20059–63.

[148] Kisseleva T, Bhattacharya S, Braunstein J, Schindler CW. Signaling through the JAK/STAT pathway, recent advances and future challenges. Gene. 2002;285(1-2):1–24.

[149] Rawlings JS, Rosler KM, Harrison DA. The JAK/STAT signaling pathway. Journal of Cell Science. 2004;117(Pt 8):1281–3.

[150] Murray PJ. The JAK-STAT signaling pathway: input and output integration. Journal of Immunology. 2007;178(5):2623–9.

[151] Darnell JE. STATs and gene regulation. Science. 1997;277(5332):1630–5.

[152] Cheng F, Wang H-W, Cuenca A, Huang M, Ghansah T, Brayer J, et al. A critical role for Stat3 signaling in immune tolerance. Immunity. 2003;19(3):425–36.

[153] Yang XO, Panopoulos AD, Nurieva R, Chang SH, Wang D, Watowich SS, et al. STAT3 regulates cytokine-mediated generation of inflammatory helper T cells. The Journal of Biological Chemistry. 2007;282(13):9358–63.

[154] Alonzi T, Maritano D, Gorgoni B, Rizzuto G, Libert C, Poli V. Essential role of STAT3 in the control of the acute-phase response as revealed by inducible gene inactivation

[correction of activation] in the liver. Molecular and Cellular Biology. 2001;21(5): 1621–32.

[155] Tian SS, Lamb P, Seidel HM, Stein RB, Rosen J. Rapid activation of the STAT3 transcription factor by granulocyte colony-stimulating factor. Blood. 1994;84(6):1760–4.

[156] Hansen ML, Woetmann A, Krejsgaard TOR, Kopp KLM, S o kilde R, Litman T, et al. IFN-alpha primes T-and NK-cells for IL-15-mediated signaling and cytotoxicity. Molecular Immunology. 2011;48(15-16):2087–93.

[157] Darnell JE, Kerr IM, Stark GR. Jak-STAT pathways and transcriptional activation in response to IFNs and other extracellular signaling proteins. Science. 1994;264(5164): 1415–21.

[158] Ruff-Jamison S, Zhong Z, Wen Z, Chen K, Darnell JE, Cohen S. Epidermal growth factor and lipopolysaccharide activate Stat3 transcription factor in mouse liver. The Journal of Biological Chemistry. 1994;269(35):21933–5.

[159] Yang J, van Oosten AL, Theunissen TW, Guo G, Silva JCR, Smith A. Stat3 activation is limiting for reprogramming to ground state pluripotency. Cell Stem Cell. 2010;7(3): 319–28.

[160] Akira S. Roles of STAT3 defined by tissue-specific gene targeting. Oncogene. 2000;19(21):2607–11.

[161] Lafarge S, Hamzeh-Cognasse H, Richard Y, Pozzetto B, Cogn e M, Cognasse F, et al. Complexes between nuclear factor-kappaB p65 and signal transducer and activator of transcription 3 are key actors in inducing activation-induced cytidine deaminase expression and immunoglobulin A production in CD40L plus interleukin-10-treated human blood B cells. Clinical and Experimental Immunology. 2011;166(2):171–83.

[162] Maritano D, Sugrue ML, Tininini S, Dewilde S, Strobl B, Fu X, et al. The STAT3 isoforms alpha and beta have unique and specific functions. Nature Immunology. 2004;5(4):401–9.

[163] Akira S. IL-6-regulated transcription factors. The International Journal of Biochemistry & Cell Biology. 1997;29(12):1401–18.

[164] Starr R, Hilton DJ. Negative regulation of the JAK/STAT pathway. BioEssays : news and reviews in molecular, cellular and developmental biology. 1999;21(1):47–52.

[165] Alexander WS. Suppressors of cytokine signalling (SOCS) in the immune system. Nature Reviews Immunology. 2002 Jun;2(6):410–6.

[166] Kershaw NJ, Murphy JM, Liau NPD, Varghese LN, Laktyushin A, Whitlock EL, et al. SOCS3 binds specific receptor-JAK complexes to control cytokine signaling by direct kinase inhibition. Nature Structural & Molecular Biology. 2013;20(4):469–76.

[167] Waiboci LW, Ahmed CM, Mujtaba MG, Flowers LO, Martin JP, Haider MI, et al. Both the suppressor of cytokine signaling 1 (SOCS-1) kinase inhibitory region and

SOCS-1 mimetic bind to JAK2 autophosphorylation site: implications for the development of a SOCS-1 antagonist. Journal of Immunology. 2007;178(8):5058–68.

[168] Jo D, Liu D, Yao S, Collins RD, Hawiger J. Intracellular protein therapy with SOCS3 inhibits inflammation and apoptosis. Nature Medicine. 2005;11(8):892–8.

[169] Yoshimura A, Naka T, Kubo M. SOCS proteins, cytokine signalling and immune regulation. Nature Reviews Immunology. 2007;7(6):454–65.

[170] Ahmed, PJAK2. Enhancement of antiviral immunity by small molecule antagonist of suppressor of cytokine signaling. Journal of Immunology. 2010;185(2):1103–13.

[171] Hankey PA. Regulation of hematopoietic cell development and function by Stat3. Frontiers in Bioscience. 2009;14:5273–90.

[172] Wang J, Rao S, Chu J, Shen X, Levasseur DN, Theunissen TW, et al. A protein interaction network for pluripotency of embryonic stem cells. Nature. 2006;444(7117):364–8.

[173] Pan G, Thomson JA. Nanog and transcriptional networks in embryonic stem cell pluripotency. Cell Research. 2007;17(1):42–9.

[174] Boyer LA, Lee TI, Cole MF, Johnstone SE, Levine SS, Zucker JP, et al. Core transcriptional regulatory circuitry in human embryonic stem cells. Cell. 2005;122(6):947–56.

[175] Li Y-Q. Master stem cell transcription factors and signaling regulation. Cellular Reprogramming. 2010;12(1):3–13.

[176] Neri F, Zippo A, Krepelova A, Cherubini A, Rocchigiani M, Oliviero S. Myc regulates the transcription of the PRC2 gene to control the expression of developmental genes in embryonic stem cells. Molecular and Cellular Biology. 2012;32(4):840–51.

[177] Goodliffe JM, Cole MD, Wieschaus E. Coordinated regulation of Myc trans-activation targets by Polycomb and the Trithorax group protein Ash1. BMC Molecular Biology. 2007;8:40.

[178] Liu S-P, Harn H-J, Chien Y-J, Chang C-H, Hsu C-Y, Fu R-H, et al. n-Butylidenephthalide (BP) Maintains Stem Cell Pluripotency by Activating Jak2/Stat3 Pathway and Increases the Efficiency of iPS Cells Generation. PloS One. 2012 Sep 7;7(9):e44024.

[179] Niwa H, Burdon T, Chambers I, Smith A. Self-renewal of pluripotent embryonic stem cells is mediated via activation of STAT3. Genes & Development. 1998;12(13):2048–60.

[180] Wei CL, Miura T, Robson P, Lim S-K, Xu X-Q, Lee MY-C, et al. Transcriptome Profiling of Human and Murine ESCs Identifies Divergent Paths Required to Maintain the Stem Cell State. Stem Cells. 2005 Feb;23(2):166–85.

[181] Vallier L. Activin/Nodal and FGF pathways cooperate to maintain pluripotency of human embryonic stem cells. Journal of Cell Science. 2005 Sep 13;118(19):4495–509.

[182] James D. TGF /activin/nodal signaling is necessary for the maintenance of pluripotency in human embryonic stem cells. Development. 2005 Feb 9;132(6):1273–82.

[183] Dahéron L, Opitz SL, Zaehres H, Lensch MW, Lensch WM, Andrews PW, et al. LIF/STAT3 signaling fails to maintain self-renewal of human embryonic stem cells. Stem Cells. 2004;22(5):770–8.

[184] Niwa H, Ogawa K, Shimosato D, Adachi K. A parallel circuit of LIF signalling pathways maintains pluripotency of mouse ES cells. Nature. 2009;460(7251):118–22.

[185] Hall J, Guo G, Wray J, Eyres I, Nichols J, Grotewold L, et al. Oct4 and LIF/Stat3 Additively Induce Krüppel Factors to Sustain Embryonic Stem Cell Self-Renewal. Cell Stem Cell. 2009 Dec;5(6):597–609.

[186] Ying Q-L, Nichols J, Chambers I, Smith A. BMP induction of Id proteins suppresses differentiation and sustains embryonic stem cell self-renewal in collaboration with STAT3. Cell. 2003 Oct 31;115(3):281–92.

[187] Armstrong L, Hughes O, Yung S, Hyslop L, Stewart R, Wappler I, et al. The role of PI3K/AKT, MAPK/ERK and NFkappabeta signalling in the maintenance of human embryonic stem cell pluripotency and viability highlighted by transcriptional profiling and functional analysis. Human Molecular Genetics. 2006;15(11):1894–913.

[188] Lee H, Herrmann A, Deng J-H, Kujawski M, Niu G, Li Z, et al. Persistently Activated Stat3 Maintains Constitutive NF-κB Activity in Tumors. Cancer Cell. 2009;15(4):283–93.

[189] Cavaleri F, Sch o ler HR. Nanog: a new recruit to the embryonic stem cell orchestra. Cell. 2003;113(5):551–2.

[190] Matsuda T, Nakamura T, Nakao K, Arai T, Katsuki M, Heike T, et al. STAT3 activation is sufficient to maintain an undifferentiated state of mouse embryonic stem cells. The EMBO Journal. 1999;18(15):4261–9.

[191] Hawkins K, Mohamet L, Ritson S, Merry CLR, Ward CM. E-cadherin and, in its absence, N-cadherin promotes Nanog expression in mouse embryonic stem cells via STAT3 phosphorylation. Stem Cells. 2012;30(9):1842–51.

[192] Do DV, Ueda J, Messerschmidt DM, Lorthongpanich C, Zhou Y, Feng B, et al. A genetic and developmental pathway from STAT3 to the OCT4-NANOG circuit is essential for maintenance of ICM lineages in vivo. Genes & Development. 2013;27(12): 1378–90.

[193] Kidder BL, Yang J, Palmer S. Stat3 and c-Myc genome-wide promoter occupancy in embryonic stem cells. PloS One. 2008;3(12):e3932.

[194] Schuringa JJ, Timmer H, Luttickhuizen D, Vellenga E, Kruijer W. c-Jun and c-Fos cooperate with STAT3 in IL-6-induced transactivation of the IL-6 respone element (IRE). Cytokine. 2001;14(2):78–87.

[195] Shaulian E, Karin M. AP-1 as a regulator of cell life and death. Nature Cell Biology. 2002;4(5):E131–6.

[196] Hu X, Paik PK, Chen J, Yarilina A, Kockeritz L, Lu TT, et al. IFN-gamma suppresses IL-10 production and synergizes with TLR2 by regulating GSK3 and CREB/AP-1 proteins. Immunity. 2006;24(5):563–74.

[197] Saeki Y, Nagashima T, Kimura S, Okada-Hatakeyama M. An ErbB receptor-mediated AP-1 regulatory network is modulated by STAT3 and c-MYC during calcium-dependent keratinocyte differentiation. Experimental Dermatology. 2012 Mar 15;21(4): 293–8.

[198] Hu G, Wade PA. NuRD and pluripotency: a complex balancing act. Cell Stem Cell. 2012;10(5):497–503.

[199] Aguilera C, Nakagawa K, Sancho R, Chakraborty A, Hendrich B, Behrens A. c-Jun N-terminal phosphorylation antagonises recruitment of the Mbd3/NuRD repressor complex. Nature. 2011;469(7329):231–5.

[200] Kortylewski M, Kujawski M, Wang T, Wei S, Zhang S, Pilon-Thomas S, et al. Inhibiting Stat3 signaling in the hematopoietic system elicits multicomponent antitumor immunity. Nature Medicine. 2005;11(12):1314–21.

[201] Welte T, Zhang SSM, Wang T, Zhang Z, Hesslein DGT, Yin Z, et al. STAT3 deletion during hematopoiesis causes Crohn's disease-like pathogenesis and lethality: a critical role of STAT3 in innate immunity. Proceedings of the National Academy of Sciences. 2003;100(4):1879–84.

[202] Schuringa JJ, Wierenga AT, Kruijer W, Vellenga E. Constitutive Stat3, Tyr705, and Ser727 phosphorylation in acute myeloid leukemia cells caused by the autocrine secretion of interleukin-6. Blood. 2000;95(12):3765–70.

[203] Gur-Wahnon D, Borovsky Z, Beyth S, Liebergall M, Rachmilewitz J. Contact-dependent induction of regulatory antigen-presenting cells by human mesenchymal stem cells is mediated via STAT3 signaling. Experimental Hematology. 2007;35(3):426–33.

[204] Phinney DG, Prockop DJ. Concise review: mesenchymal stem/multipotent stromal cells: the state of transdifferentiation and modes of tissue repair--current views. Stem cells. 2007;25(11):2896–902.

[205] Chamberlain G, Fox J, Ashton B, Middleton J. Concise review: mesenchymal stem cells: their phenotype, differentiation capacity, immunological features, and potential for homing. Stem Cells. 2007;25(11):2739–49.

[206] Uccelli A, Moretta L, Pistoia V. Mesenchymal stem cells in health and disease. Nature Reviews Immunology. 2008;8(9):726–36.

[207] Martinez-Outschoorn UE, Balliet RM, Lin Z, Whitaker-Menezes D, Howell A, Sotgia F, et al. Hereditary ovarian cancer and two-compartment tumor metabolism: epithe-

lial loss of BRCA1 induces hydrogen peroxide production, driving oxidative stress and NFkappaB activation in the tumor stroma. Cell Cycle. 2012;11(22):4152–66.

[208] Uchibori R, Tsukahara T, Mizuguchi H, Saga Y, Urabe M, Mizukami H, et al. NF-kappaB activity regulates mesenchymal stem cell accumulation at tumor sites. Cancer Research. 2013;73(1):364–72.

[209] Vultur A, Cao J, Arulanandam R, Turkson J, Jove R, Greer P, et al. Cell-to-cell adhesion modulates Stat3 activity in normal and breast carcinoma cells. Oncogene. 2004;23(15):2600–16.

[210] Oellerich T, Oellerich MF, Engelke M, Munch S, Mohr S, Nimz M, et al. 2 integrin-derived signals induce cell survival and proliferation of AML blasts by activating a Syk/STAT signaling axis. Blood. 2013;121(19):3889–99.

[211] Gough DJ, Corlett A, Schlessinger K, Wegrzyn J, Larner AC, Levy DE. Mitochondrial STAT3 supports Ras-dependent oncogenic transformation. Science. 2009;324(5935): 1713–6.

[212] Bromberg JF, Horvath CM, Besser D, Lathem WW, Darnell JE. Stat3 activation is required for cellular transformation by v-src. Molecular and Cellular Biology. 1998;18(5):2553–8.

[213] Bromberg J, Wang TC. Inflammation and Cancer:IL-6 and STAT3 Complete the Link. Cancer cell. 2009 Feb 3;15(2):79–80.

[214] Grivennikov S, Karin M. Autocrine IL-6 signaling: a key event in tumorigenesis? Cancer Cell. 2008;13(1):7–9.

[215] Gur-Wahnon D, Borovsky Z, Liebergall M, Rachmilewitz J. The Induction of APC with a Distinct Tolerogenic Phenotype via Contact-Dependent STAT3 Activation. PloS One. 2009;4(8):e6846.

[216] Bournazou E, Bromberg J. Targeting the tumor microenvironment: JAK-STAT3 signaling. JAK-STAT. 2013;2(2):e23828.

[217] Shain KH, Yarde DN, Meads MB, Huang M, Jove R, Hazlehurst LA, et al. 1 Integrin Adhesion Enhances IL-6-Mediated STAT3 Signaling in Myeloma Cells: Implications for Microenvironment Influence on Tumor Survival and Proliferation. Cancer Research. 2009 Jan 20;69(3):1009–15.

[218] Yu H, Kortylewski M, Pardoll D. Crosstalk between cancer and immune cells: role of STAT3 in the tumour microenvironment. Nature Rev Immunol. 2007 Jan;7(1):41–51.

[219] Dreesen O, Brivanlou AH. Signaling pathways in cancer and embryonic stem cells. Stem Cell Reviews. 2007;3(1):7–17.

[220] Kim J, Woo AJ, Chu J, Snow JW, Fujiwara Y, Kim CG, et al. A Myc Network Accounts for Similarities between Embryonic Stem and Cancer Cell Transcription Programs. Cell. 2010;143(2):313–24.

[221] Huntly BJP, Gilliland DG. Leukaemia stem cells and the evolution of cancer-stem-cell research. Nature Reviews Cancer. 2005;5(4):311–21.

[222] Reya T, Morrison SJ, Clarke MF, Weissman IL. Stem cells, cancer, and cancer stem cells. Nature. 2001;414(6859):105–11.

[223] Pardal R, Clarke MF, Morrison SJ. Applying the principles of stem-cell biology to cancer. Nature Reviews Cancer. 2003;3(12):895–902.

[224] Bjerkvig R, Tysnes BB, Aboody KS, Najbauer J, Terzis AJA. Opinion: the origin of the cancer stem cell: current controversies and new insights. Nature Reviews Cancer. 2005;5(11):899–904.

[225] Agarwal JR, Matsui W. Multiple myeloma: a paradigm for translation of the cancer stem cell hypothesis. Anti-cancer agents in medicinal chemistry. 2010;10(2):116–20.

[226] Frank NY, Schatton T, Frank MH. The therapeutic promise of the cancer stem cell concept. The Journal of clinical investigation. 2010;120(1):41–50.

[227] Jordan CT, Guzman ML, Noble M. Cancer stem cells. N Engl J Med. 2006;355(12): 1253–61.

[228] Lin C, Wang L, Wang H, Yang L, Guo H, Wang X. Tanshinone IIA inhibits breast cancer stem cells growth in vitro and in vivo through attenuation of IL-6/STAT3/NF-kB signaling pathways. J Cell Biochem. 2013 Jul 18;114(9):2061–70.

[229] Iliopoulos D, Hirsch HA, Struhl K. An Epigenetic Switch Involving NF-κB, Lin28, Let-7 MicroRNA, and IL6 Links Inflammation to Cell Transformation. Cell. 2009;139(4):693–706.

[230] Glinka Y, Mohammed N, Subramaniam V, Jothy S, Prud'homme GERJ. Neuropilin-1 is expressed by breast cancer stem-like cells and is linked to NF-kappaB activation and tumor sphere formation. Biochemical and Biophysical Research Communications. 2012;425(4):775–80.

[231] Long H, Xie R, Xiang T, Zhao Z, Lin S, Liang Z, et al. Autocrine CCL5 signaling promotes invasion and migration of CD133+ovarian cancer stem-like cells via NF-kappaB-mediated MMP-9 upregulation. Stem Cells. 2012;30(10):2309–19.

[232] Inoue J-I, Gohda J, Akiyama T, Semba K. NF-kappaB activation in development and progression of cancer. Cancer Science. 2007;98(3):268–74.

[233] Yang J, van Oosten AL, Theunissen TW, Guo G, Silva JCR, Smith A. Stat3 Activation Is Limiting for Reprogramming to Ground State Pluripotency. Cell Stem Cell. 2010;7(3):319–28.

[234] van Oosten AL, Costa Y, Smith A, Silva JECR. JAK/STAT3 signalling is sufficient and dominant over antagonistic cues for the establishment of naive pluripotency. Nature Communications. 2012;3:817.

[235] Nichols J, Silva J, Roode M, Smith A. Suppression of Erk signalling promotes ground state pluripotency in the mouse embryo. Development. 2009;136(19):3215–22.

[236] Ura H, Usuda M, Kinoshita K, Sun C, Mori K, Akagi T, et al. STAT3 and Oct-3/4 Control Histone Modification through Induction of Eed in Embryonic Stem Cells. The Journal of Biological Chemistry. 2008;283(15):9713–23.

[237] Kim J, Chu J, Shen X, Wang J, Orkin SH. An Extended Transcriptional Network for Pluripotency of Embryonic Stem Cells. Cell. 2008;132(6):1049–61.

[238] Orkin SH, Hochedlinger K. Chromatin Connections to Pluripotency and Cellular Reprogramming. Cell. 2011;145(6):835–50.

[239] Tang Y, Luo Y, Jiang Z, Ma Y, Lin C-J, Kim C, et al. Jak/Stat3 Signaling Promotes Somatic Cell Reprogramming by Epigenetic Regulation. Stem Cells. 2012;30(12):2645–56.

[240] Icardi L, De Bosscher K, Tavernier J. Cytokine & Growth Factor Reviews. Cytokine & Growth Factor Reviews. 2012;23(6):283–91.

[241] Huangfu D, Osafune K, Maehr RE, Guo W, Eijkelenboom A, Chen S, et al. Induction of pluripotent stem cells from primary human fibroblasts with only Oct4 and Sox2. Nature Biotechnology. 2008;26(11):1269–75.

[242] Ang Y-S, Gaspar-Maia A, Lemischka IR, Bernstein E. Stem cells and reprogramming: breaking the epigenetic barrier? Trends in Pharmacological Sciences. 2011;32(7):394–401.

[243] Wu H, Zhang Y. Mechanisms and functions of Tet protein-mediated 5-methylcytosine oxidation. Genes & Development. 2011;25(23):2436–52.

[244] Tan L, Shi YG. Tet family proteins and 5-hydroxymethylcytosine in development and disease. Development. 2012;139(11):1895–902.

[245] Branco MR, Ficz G, Reik W. Uncovering the role of 5-hydroxymethylcytosine in the epigenome. Nature reviews Genetics. 2011 Nov 15;13(1):7–13.

[246] Cimmino L, Abdel-Wahab O, Levine RL, Aifantis I. TET family proteins and their role in stem cell differentiation and transformation. Cell Stem Cell. 2011;9(3):193–204.

[247] Fan Y, Mao R, Yang J. NF-kappaB and STAT3 signaling pathways collaboratively link inflammation to cancer. Protein & Cell. 2013;4(3):176–85.

[248] Gough DJ, Levy DE, Johnstone RW, Clarke CJ. IFNγ signaling—Does it mean JAK–STAT? Cytokine & Growth Factor Reviews. 2008;19(5-6):383–94.

[249] Grivennikov SI, Karin M. Cytokine & Growth Factor Reviews. Cytokine & Growth Factor Reviews. 2010;21(1):11–9.

[250] Brasier AR. The nuclear factor-kappaB-interleukin-6 signalling pathway mediating vascular inflammation. Cardiovascular Research. 2010;86(2):211–8.

[251] Wang R. Activation of Stat3 Sequence-specific DNA Binding and Transcription by p300/CREB-binding Protein-mediated Acetylation. The Journal of Biological Chemistry. 2005;280(12):11528–34.

[252] Nadiminty N, Lou W, Lee SO, Lin X, Trump DL, Gao AC. Stat3 activation of NF-\kappa\B p100 processing involves CBP/p300-mediated acetylation. Proceedings of the National Academy of Sciences. 2006;103(19):7264–9.

[253] Schmidlin H, Diehl SA, Blom B. New insights into the regulation of human B-cell differentiation. Trends in Immunology. 2009;30(6):277–85.

[254] Krutzik SR, Tan B, Li H, Ochoa MT, Liu PT, Sharfstein SE, et al. TLR activation triggers the rapid differentiation of monocytes into macrophages and dendritic cells. Nature Medicine. 2005;11(6):653–60.

[255] Hu X, Herrero C, Li W-P, Antoniv TT, Falck-Pedersen E, Koch AE, et al. Sensitization of IFN-gamma Jak-STAT signaling during macrophage activation. Nature Immunology. 2002;3(9):859–66.

[256] Ivashkiv LB. A signal-switch hypothesis for cross-regulation of cytokine and TLR signalling pathways. Nature Reviews Immunology. 2008;8(10):816–22.

[257] Wang L, Gordon RA, Huynh L, Su X, Park Min K-H, Han J, et al. Indirect inhibition of Toll-like receptor and type I interferon responses by ITAM-coupled receptors and integrins. Immunity. 2010;32(4):518–30.

[258] Hu X, Chen J, Wang L, Ivashkiv LB. Crosstalk among Jak-STAT, Toll-like receptor, and ITAM-dependent pathways in macrophage activation. Journal of Leukocyte Biology. 2007;82(2):237–43.

[259] Chen X, Xu H, Yuan P, Fang F, Huss M, Vega VB, et al. Integration of external signaling pathways with the core transcriptional network in embryonic stem cells. Cell. 2008;133(6):1106–17.

[260] Schugar RC, Robbins PD, Deasy BM. Small molecules in stem cell self-renewal and differentiation. Gene Therapy. 2008;15(2):126–35.

[261] Hirai H, Karian P, Kikyo N. Regulation of embryonic stem cell self-renewal and pluripotency by leukaemia inhibitory factor. Biochem J. 2011;438(1):11–23.

[262] Richards M, Tan S-P, Tan J-H, Chan W-K, Bongso A. The transcriptome profile of human embryonic stem cells as defined by SAGE. Stem Cells. 2004;22(1):51–64.

[263] Melton C, Judson RL, Blelloch R. Opposing microRNA families regulate self-renewal in mouse embryonic stem cells. Nature. 2010;463(7281):621–6.

[264] Yu J, Vodyanik MA, Smuga-Otto K, Antosiewicz-Bourget J, Frane JL, Tian S, et al. Induced pluripotent stem cell lines derived from human somatic cells. Science. 2007;318(5858):1917–20.

[265] Darr H, Benvenisty N. Genetic analysis of the role of the reprogramming gene LIN-28 in human embryonic stem cells. Stem Cells. 2009;27(2):352–62.

[266] Koche RP, Smith ZD, Adli M, Gu H, Ku M, Gnirke A, et al. Reprogramming Factor Expression Initiates Widespread Targeted Chromatin Remodeling. Cell Stem Cell. 2011;8(1):96–105.

[267] Papp B, Plath K. Epigenetics of Reprogramming to Induced Pluripotency. Cell. 2013;152(6):1324–43.

[268] Reynolds N, Latos P, Hynes-Allen A, Loos R, Leaford D, O'Shaughnessy A, et al. NuRD suppresses pluripotency gene expression to promote transcriptional heterogeneity and lineage commitment. Cell Stem Cell. 2012;10(5):583–94.

[269] Zhu D, Fang J, Li Y, Zhang J. Mbd3, a Component of NuRD/Mi-2 Complex, Helps Maintain Pluripotency of Mouse Embryonic Stem Cells by Repressing Trophectoderm Differentiation. Zwaka T, editor. PloS One. 2009;4(11):e7684.

[270] Morey L, Brenner C, Fazi F, Villa R, Gutierrez A, Buschbeck M, et al. MBD3, a Component of the NuRD Complex, Facilitates Chromatin Alteration and Deposition of Epigenetic Marks. Molecular and Cellular Biology. 2008 Sep;28(19):5912–23.

[271] Sakai H. MBD3 and HDAC1, Two Components of the NuRD Complex, Are Localized at Aurora-A-positive Centrosomes in M Phase. Journal of Biological Chemistry. 2002;277(50):48714–23.

[272] Gunther K, Rust M, Leers J, Boettger T, Scharfe M, Jarek M, et al. Differential roles for MBD2 and MBD3 at methylated CpG islands, active promoters and binding to exon sequences. Nucleic Acids Research. 2013;41(5):3010–21.

[273] Kaji K, Nichols J, Hendrich B. Mbd3, a component of the NuRD co-repressor complex, is required for development of pluripotent cells. Development. 2007;134(6): 1123–32.

[274] Luo M, Ling T, Xie W, Sun H, Zhou Y, Zhu Q, et al. NuRD Blocks Reprogramming of Mouse Somatic Cells into Pluripotent Stem Cells. Stem Cells. 2013;31(7):1278–86.

[275] Rais Y, Zviran A, Geula S, Gafni O, Chomsky E, Viukov S, et al. Deterministic direct reprogramming of somatic cells to pluripotency. Nature. 2013;502(7469):65–70.

[276] Silva J, Smith A. Capturing Pluripotency. Cell. 2008;132(4):532–6.

[277] Xu N, Papagiannakopoulos T, Pan G, Thomson JA, Kosik KS. MicroRNA-145 regulates OCT4, SOX2, and KLF4 and represses pluripotency in human embryonic stem cells. Cell. 2009;137(4):647–58.

[278] Fazzio TG, Rando OJ. NURDs are required for diversity. The EMBO journal. 2012;31(14):3036–7.

[279] Bernstein BE, Mikkelsen TS, Xie X, Kamal M, Huebert DJ, Cuff J, et al. A Bivalent Chromatin Structure Marks Key Developmental Genes in Embryonic Stem Cells. Cell. 2006;125(2):315–26.

[280] Gan Q, Yoshida T, McDonald OG, Owens GK. Concise review: epigenetic mechanisms contribute to pluripotency and cell lineage determination of embryonic stem cells. Stem Cells. 2007;25(1):2–9.

[281] Novershtern N, Hanna JH. esBAF safeguards Stat3 binding to maintain pluripotency. Nature Cell Biology. 2011;13(8):886–8.

[282] Liang G, Zhang Y. Embryonic stem cell and induced pluripotent stem cell: an epigenetic perspective. Cell Research. 2012;23(1):49–69.

[283] Araki Y, Wang Z, Zang C, Wood WH III, Schones D, Cui K, et al. Genome-wide Analysis of Histone Methylation Reveals Chromatin State-Based Regulation of Gene Transcription and Function of Memory CD8. Immunity. 2009;30(6):912–25.

[284] Ku M, Koche RP, Rheinbay E, Mendenhall EM, Endoh M, Mikkelsen TS, et al. Genomewide analysis of PRC1 and PRC2 occupancy identifies two classes of bivalent domains. PLoS Genetics. 2008;4(10):e1000242.

[285] Li M, Liu G-H, Izpisua Belmonte JC. Navigating the epigenetic landscape of pluripotent stem cells. Nature Reviews Molecular Cell Biology. 2012;13(8):524–35.

[286] Cui Y, Cho I-H, Chowdhury B, Irudayaraj J. Real-time dynamics of methyl-CpG-binding domain protein 3 and its role in DNA demethylation by fluorescence correlation spectroscopy. Epigenetics. 2013;8(10):10–9.

[287] Shen L, Zhang Y. 5-Hydroxymethylcytosine: generation, fate, and genomic distribution. Current Opinion in Cell Biology. 2013;25(3):289–96.

[288] Xu Y, Wu F, Tan L, Kong L, Xiong L, Deng J, et al. Genome-wide Regulation of 5hmC, 5mC, and Gene Expression by Tet1 Hydroxylase in Mouse Embryonic Stem Cells. Molecular cell. 2011;42(4):451–64.

[289] Yildirim O, Li R, Hung J-H, Chen PB, Dong X, Ee L-S, et al. Mbd3/NURD complex regulates expression of 5-hydroxymethylcytosine marked genes in embryonic stem cells. Cell. 2011;147(7):1498–510.

[290] Liang J, Wan M, Zhang Y, Gu P, Xin H, Jung SY, et al. Nanog and Oct4 associate with unique transcriptional repression complexes in embryonic stem cells. Nature Cell Biology. 2008;10(6):731–9.

[291] Luo M, Ling T, Xie W, Sun H, Zhou Y, Zhu Q, et al. NuRD blocks reprogramming of mouse somatic cells into pluripotent stem cells. Stem Cells. 2013;31(7):1278–86.

[292] Hu G, Wade PA. NuRD and Pluripotency: A Complex Balancing Act. Cell Stem Cell. 2012;10(5):497–503.

[293] Kaji K, Caballero IMIN, MacLeod R, Nichols J, Wilson VA, Hendrich B. The NuRD component Mbd3 is required for pluripotency of embryonic stem cells. Nature Cell Biology. 2006;8(3):285–92.

[294] Fan Y, Mao R, Yang J. NF-κB and STAT3 signaling pathways collaboratively link inflammation to cancer. Protein & Cell. 2013;4(3):176–85.

[295] Koche RP, Smith ZD, Adli M, Gu H, Ku M, Gnirke A, et al. Reprogramming factor expression initiates widespread targeted chromatin remodeling. Cell Stem Cell. 2011;8(1):96–105.

[296] Chen X, Gazzar El M, Yoza BK, McCall CE. The NF-B Factor RelB and Histone H3 Lysine Methyltransferase G9a Directly Interact to Generate Epigenetic Silencing in Endotoxin Tolerance. The Journal of Biological Chemistry. 2009;284(41):27857–65.

Chapter 2

Embryonic Stem Cells and Oncogenes

Hiroshi Koide

Additional information is available at the end of the chapter

http://dx.doi.org/10.5772/57614

1. Introduction

Stem cells possess two main attributes, namely, a self-renewal capacity and multipotency. Self-renewal refers to the ability of a cell to replicate itself without differentiating or losing multipotency. Multipotency is the ability to differentiate into more than one cell lineage. Recent studies have revealed that there are several types of stem cells. For example, neural stem cells can differentiate into neurons, astrocytes, and oligodendrocytes, while hematopoietic stem cells can differentiate into all types of blood cells. Embryonic stem (ES) cells were originally established from the inner cell mass (ICM) of mouse blastocysts [1, 2]. As the ICM gives rise to the embryo, ES cells can differentiate into most cell types in the body and are therefore pluripotent. The successful establishment of human pluripotent cells, namely, human ES cells and human induced pluripotent (iPS) cells, opens up the possibility of using these cells in regenerative medicine [3-5]. However, several issues remain to be resolved. One such issue is unanticipated tumor formation by transplanted pluripotent cells, which seems to be associated with the self-renewal capacity of these cells.

The self-renewal of mouse ES cells [6, 7] can be maintained by the presence of leukemia inhibitory factor (LIF) in the culture medium. LIF binds to LIF receptor (LIFR) and induces formation of gp130/LIFR heterodimers, leading to activation of the downstream transcription factor signal transducer and activator of transcription (STAT)-3. The LIF/STAT3 pathway plays a critical role in ES cell self-renewal by upregulating several self-renewal genes [8]. The expression of a set of self-renewal genes is also regulated by networks of important transcription factors, including Oct3/4, Sox2, and Nanog [9].

Cancer is one of the most feared diseases throughout the world. Approximately 10 million people die from cancer every year and the number of cancer patients is increasing. This disease is characterized by loss of cellular growth control, which is induced by genome alterations, such as DNA sequence changes, copy number aberrations, chromosomal rearrangements, and modifications in DNA methylation. Recent studies have provided evidence that tumor tissue contains a small subset of stem-like cells called cancer stem cells. Cancer stem cells have stem cell-like attributes, namely, self-renewal and differentiation, which enables them to produce tumors by self-renewing and giving rise to differentiated progeny. The concept of cancer stem cells first arose from studies of leukemia stem cells [10] and subsequently of solid tumors [11, 12]. Given that cancer stem cells play a prominent role in cancer cell growth, it may be reasonable to expect similarities between ES cells and cancer cells.

Indeed, ES cells are similar to cancer cells in several respects. When injected into immunodeficient mice, ES cells and cancer cells can produce benign and malignant tumors, respectively. Both cell types have a rapid cell cycle, which results in fast proliferation. Telomerase activity is very high in both cell types, which allows them to proliferate indefinitely. Both cell types contain a "side population of cells" with a high drug efflux capacity, which gives rise to their drug-resistance phenotype. In addition, several signal transduction pathways seem to be used in both ES cell self-renewal and cancer cell growth. For example, the STAT3 pathway, which plays a central role in ES cell self-renewal, is activated in several types of cancer cells. On the other hand, the Wnt/β-catenin pathway, whose activation is associated with tumorigenesis in many tissues, is involved in ES cell self-renewal. Moreover, poorly differentiated tumors preferentially overexpress genes that are normally enriched in ES cells [13].

The similarities between ES cells and cancer cells raise the possibility that certain molecules that are involved in ES cell self-renewal play important roles in cancer cell growth, while certain oncogenes play critical roles in ES cell self-renewal (Fig. 1). In this chapter, I will provide examples of such molecules and describe their roles in ES cell self-renewal and tumorigenesis.

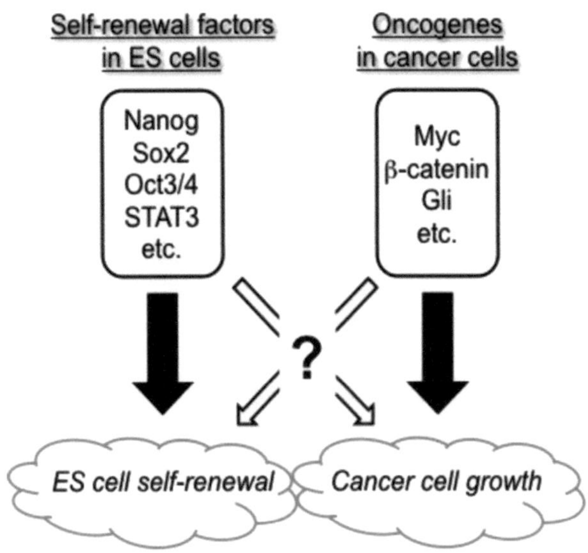

Figure 1. Similarity between ES cells and cancer cells.

2. ES cell self-renewal genes that are involved in cancer cell growth

2.1. Oct3/4

Oct3/4 was originally isolated by three groups [14-16]. Since Oct1 and Oct2 had already been identified, two groups named this protein Oct3, while the other group named it Oct4. Therefore, I will use the term "Oct3/4" to describe this protein. Oct3/4 is a transcription factor belonging to the POU family and is encoded by the *pou5f1* gene. The Oct3/4 protein contains three domains, namely, the N-terminal, POU, and C-terminal domains. The N- and C-terminal domains are transactivation domains with redundant functions, while the POU domain is a bipartite DNA-binding domain consisting of the POU-specific domain and the POU homeodomain. In mouse development, deficiency of this transcription factor results in loss of the ICM [17]. In agreement with this, conditional knockout of this gene in ES cells results in their differentiation into trophectoderm cells [18], indicating that Oct3/4 is a central player in the self-renewal of ES cells. Furthermore, the recent finding that Oct3/4 is one of the four factors required for the production of iPS cells indicates the importance of Oct3/4 for the acquisition of pluripotency [19].

Not only suppression, but also overexpression of Oct3/4 induces ES cell differentiation [18], suggesting that the activity of Oct3/4 needs to be sustained at the correct level to maintain ES cell self-renewal. Oct3/4 expression in ES cells is positively and negatively regulated by multiple factors. The upstream region of the *pou5f1* gene contains proximal and distal enhancers, which regulate stem cell-specific expression of Oct3/4 [20]. An orphan nuclear receptor Lrh1 (also known as Nr5a2) binds to the proximal enhancer to maintain Oct3/4 expression [21], whereas Oct3/4 and Sox2 associate with the distal enhancer to stimulate Oct3/4 expression [22]. Negative regulators of Oct3/4 expression include Gcnf, Coup-tfs, and Cdx2, whose expression are induced upon ES cell differentiation. In addition to regulation of Oct3/4 expression, the transcriptional activity of Oct3/4 protein is regulated by Oct3/4-binding proteins. For example, it is well-established that Sox2 is a co-factor of Oct3/4. β-catenin binds to Oct3/4 and functions as a co-activator that enhances the transcriptional activity of Oct3/4 [23]. The orphan nuclear receptor Dax1 also binds to Oct3/4, but acts as a negative regulator by interfering with the binding of Oct3/4 to DNA [24].

In adult human tissues, expression of Oct3/4 is restricted to germline cells and is very low in other tissues. By contrast, Oct3/4 is expressed in several types of human cancers, including prostate, breast, oral, bladder, and seminoma [25], suggesting the importance of this transcription factor in cancer development. Moreover, Oct3/4 is highly expressed in cancer stem-like cells in breast cancer, lung cancer, and bone sarcoma [26-28]. In lung cancer, Oct3/4 maintains the properties of cancer stem-like cells [28]. In agreement with these findings, high expression of Oct3/4 significantly correlates with poor overall survival of nasopharyngeal carcinoma patients [29].

2.2. Sox2

Sox2 is a transcription factor that belongs to the SRY-related HMG-box protein (Sox) family. The Sox2 protein contains one HMG box, as well as a transactivation domain in its C-terminus. Expression of Sox2 in ES cells is mainly regulated by Oct3/4 and Sox2 itself. In addition, microRNAs (miRNAs) are involved in the control of Sox2 expression. For example, miR-9 binds to the 3′-untranslated region of *Sox2* mRNA and represses Sox2 expression [30]. In addition to regulation at the transcriptional level, the activity of Sox2 protein is regulated by post-translational modifications. Mouse Sox2 protein can be sumoylated at Lys-247, which impairs its binding to the *Fgf4* enhancer [31]. Additionally, acetylation of Lys-75 by p300/CBP promotes the nuclear export of Sox2 [32]. Akt directly interacts with Sox2 to phosphorylate Thr-118, leading to stabilization of the Sox2 protein [33].

Sox2-deficient mouse embryos die shortly after implantation [34], indicating that Sox2 is required for early development. In ES cells, Sox2 usually cooperates with Oct3/4 to regulate the expression of several self-renewal factors, including Nanog, and disruption of the *sox2* gene results in the differentiation of ES cells into trophectoderm-like cells [35]. In addition, Sox2 regulates expression of histone acetyltransferases, including Tip60 and Elp3, and forms a positive feedback loop with the polycomb group protein Eed to control the levels of histone acetylation and methylation [36]. Similar to Oct3/4, Sox2 is one of the four factors required for iPS cell production [19].

Besides its pivotal role in maintaining ES cell self-renewal, Sox2 is closely associated with many types of cancer [37]. The *Sox2* gene is located at chromosome 3q26, a region that is frequently amplified in carcinomas. Amplification of the *Sox2* gene has been observed in human squamous cell carcinomas of the lung and esophagus [38]. Sox2 is involved in the tumorigenesis of several types of tumors, such as lung, breast, skin, prostate, ovarian, and sinonasal. For example, Sox2 is expressed in early-stage breast tumors, and high Sox2 expression is associated with large tumor size [39, 40]. Sox2 expression in breast cancer enhances cancer stem cell-like properties [40]. Sox2 is also reportedly involved in regulation of cancer stem-like cells in ovarian carcinoma [41].

2.3. Nanog

Nanog was originally identified as a gene whose overexpression can bypass the LIF requirement of mouse ES cells for self-renewal [42, 43]. Nanog protein can be ubiquitinated at its PEST domain, resulting in its degradation through the proteasome pathway in ES cells [44]. Phosphorylation of Nanog prevents this ubiquitination, thereby increasing the stability of Nanog [45].

Although *Nanog*-deficient ES cells expand at a slower rate than wild-type cells, they can self-renew and retain expression of Oct3/4 and Sox2 [46]. These observations suggest that Nanog is involved in the growth of ES cells, but is dispensable for ES cell self-renewal. Human Nanog is a reprogramming factor that can produce human iPS cells [5]. In murine cells, mouse Nanog accelerates reprogramming and promotes the transition of pre-iPS cells into mature iPS cells [47].

Normally, Nanog is expressed at early embryonic stages and in germline stem cells, but not in adult tissues. However, Nanog is expressed at high levels in several types of cancers, including breast, cervical, oral and kidney [48]. Nanog is also highly expressed in germ cell tumors [49], which are characterized by the gain of the short arm of chromosome 12, at which the *Nanog* gene is located. In addition, several reports have suggested that Nanog is involved in the epithelial-mesenchymal transition and metastasis. For example, Nanog stimulates cell migration and invasion in ovarian cancer through downregulation of *E-cadherin* and *Foxj1* [50]. Nanog expression is higher in hepatocellular carcinoma cell lines that are highly metastatic than in those that are lowly metastatic, and the Nodal/Smad3 pathway plays an important role in the Nanog-stimulated epithelial-mesenchymal transition [51]. More importantly, Nanog overexpression is associated with poor prognosis in some types of cancer, such as colorectal, ovarian, and breast [48]

Nanog has 11 highly homologous pseudogenes in human cells. Of these, *NanogP8* encodes a full-length protein with only one amino acid difference from Nanog, and is involved in tumorigenesis [52, 53]. It was recently shown that NanogP8 can function as a reprogramming factor, with similar activity to Nanog [54].

2.4. STAT3

STAT3 is a downstream transcription factor of gp130, and is therefore activated by several cytokines, including interleukin (IL)-6, IL-11, and LIF. This transcription factor was initially identified as APRF (acute phase response factor), an inducible DNA-binding protein that binds to the IL-6 response element within the promoters of genes that encode hepatic acute phase proteins [55]. Human and mouse STAT3 proteins both have 770 amino acid residues and are highly homologous to each other (only three amino acid differences). In addition to the DNA-binding domain, STAT3 protein contains a SH2 domain, which facilitates dimer formation upon phosphorylation by upstream kinases including JAK2. STAT3 protein is usually in an inactive form and is localized in the cytoplasm. Upon cytokine stimulation, STAT3 is tyrosine-phosphorylated by activated JAKs. Thereafter, the phosphorylated STAT3 protein forms a homodimer or a heterodimer and translocates into the nucleus, where it stimulates the transcription of its target genes. In mouse development, STAT3 activity is detected during early post-implantation development [56] and *Stat3*-deficient mice die prior to gastrulation [57], suggesting that STAT3 plays an important role in early embryogenesis.

STAT3 plays an indispensable role in the self-renewal of mouse ES cells, and deficiency of STAT3 leads to the differentiation of these cells. By contrast, human ES cells do not require STAT3 activation for their self-renewal. This discrepancy in the requirement for STAT3 is most likely due to differences between mouse and human ES cells, as mouse ES cells are derived from the ICM, whereas human ES cells are derived from the epiblast. Extensive studies have identified many self-renewal factors that are downstream of STAT3 signaling, such as transcription factors, epigenetic regulators, and oncogenes [8].

Although the STAT3 protein is normally only activated in response to signals that control cell growth, overactivation of STAT3 protein has been observed in several types of cancer, including breast, prostate, and pancreas, as well as leukemia and lymphoma [58]. Indeed,

STAT3 is constitutively activated in nearly 70% of tumors. Because of the high frequency of its overactivation, STAT3 is considered to be a valuable target for anti-cancer therapy.

2.5. Krüppel-like factor (Klf) 4

Klfs are a family of transcription factors that play important roles in many fundamental biological processes. They were named "Krüppel-like" owing to their strong homology with the *Drosophila* gene product Krüppel, which is involved in segmentation of the developing embryo. Klf family proteins contain three C2H2-type zinc fingers that bind DNA. *Klf4* was independently cloned by two groups and named "gut-enriched KLF" and "epithelial zinc finger" owing to its high expression in the intestine and skin epithelium, respectively [59, 60]. However, it was later discovered that this transcription factor is expressed in several other tissues, such as lung, testis, and thymus. In addition to a C-terminal DNA-binding domain consisting of zinc fingers, Klf4 protein contains an activation domain in its N-terminus and a repressive domain in its central region. Probably owing to this structure, Klf4 is a bi-functional transcription factor that can either activate or repress transcription of its target genes. Similar to Oct3/4 and Sox2, Klf4 is one of the four factors that induce reprogramming of murine cells [19].

Klf4 is highly expressed in self-renewing ES cells, but not in differentiated ES cells. Klf4 regulates self-renewal-specific expression of Lefty1, in cooperation with Oct3/4 and Sox2 [61]. Klf4 is also involved in Oct3/4 expression [62]. Overexpression of Klf4 results in the inhibition of ES cell differentiation, possibly through upregulation of *Nanog* [63, 64]. In addition, there is a marked overlap between genes that are regulated by Nanog and those that are regulated by Klf4. These observations suggest the importance of Klf4 in ES cell self-renewal. However, *Klf4*-null mice have no detectable defects during embryogenesis [65]. Furthermore, a recent study reported that the function of Klf4 in ES cell self-renewal is partially redundant because combined knockdown of *Klf4*, *Klf2*, and *Klf5*, but not any one gene individually, results in spontaneous ES cell differentiation [62], suggesting that Klf4 is dispensable for ES cell self-renewal.

In cancer cells, Klf4 acts as a tumor suppressor or an oncogene, possibly owing to its bi-functionality. Whether Klf4 acts as a tumor suppressor or an oncogene likely depends on the tumor type. For example, Klf4 functions as a tumor suppressor in the intestinal and gastric epithelium, and expression of *Klf4* is downregulated in human colorectal and gastric carcinomas [66, 67]. On the other hand, overexpression of Klf4 in the skin results in squamous epithelial dysplasia, eventually leading to squamous cell carcinoma [68, 69]. A high expression level of Klf4 significantly correlates with a poor prognosis in hepatocellular carcinoma [70].

2.6. Zinc-finger protein (Zfp)-57

Zfp57 is a transcription factor that was originally identified as an undifferentiated cell-specific gene in F9 embryonal carcinoma cells [71]. Mouse Zfp57 protein contains one Kruppel-associated box (KRAB) domain and five zinc fingers, while human Zfp57 protein has one KRAB domain and seven zinc fingers. In adult mouse, Zfp57 is highly expressed in testis and brain [71, 72]. Loss of the zygotic function of Zfp57 leads to partial lethality, while eliminating both the maternal and zygotic functions of Zfp57 results in complete embryonic lethality [73]. Through its KRAB domain, Zfp57 interacts with KRAB-associated protein 1 (Kap1), a scaffold protein for heterochromatin-inducing factors, and thus participates in genome imprinting by recruiting Kap1 to multiple imprinting control regions [73, 74]. Mutations in the *Zfp57* gene cause transient neonatal diabetes mellitus type 1 [75]. Zfp57 is a downstream molecule of STAT3 and Oct3/4 in ES cells, and is therefore specifically expressed in self-renewing ES cells [76]. Zfp57 deficiency has no effect on the self-renewal or growth in ES cells, suggesting that this transcription factor is dispensable for ES cell self-renewal.

Based on our prediction that a molecule expressed in self-renewing ES cells may play an important role in cancer cell growth, we recently screened several ES cell-specific transcription factors for their tumor-promoting activity, and found that Zfp57 can promote anchorage-independent growth of human fibrosarcoma HT1080 cells [77]. Zfp57 overexpression enhances, while its knockdown suppresses, HT1080 tumor formation in nude mice. Zfp57 regulates the expression of insulin-like growth factor 2, which plays a critical role in Zfp57-induced anchorage-independent growth and tumor formation. Furthermore, overexpression of Zfp57 causes anchorage-independent growth of the mouse immortal fibroblast cell line NIH3T3, and immunohistochemical analysis revealed the overexpression of Zfp57 in several cancers, including pancreatic, gastric, breast, colon, and esophageal. These results suggest that *Zfp57* is an oncogene in some types of cancer. Moreover, we also found that Zfp57 is involved in anchorage-independent growth of ES cells and that *Zfp57*-null ES cells form smaller teratomas than the parental ES cells in immunodeficient mice, suggesting the importance of Zfp57 in teratoma formation by ES cells.

3. Oncogenes that are involved in ES cell self-renewal

3.1. β-catenin

The importance of the Wnt pathway in tumorigenesis was recognized by identification of adenomatous polyposis coli (APC) mutations in familial adenomatous polyposis [78, 79]. In Wnt signaling, the Apc protein functions as a negative regulator and is involved in degradation of β-catenin, the mammalian homologue of *Drosophila* Armadillo. β-catenin contains multiple armadillo repeats in its central region and a transcriptional activator domain in its C-terminal region. Human and mouse β-catenin proteins both have 781 amino acid residues and are almost identical to each other (only one amino acid difference). β-catenin acts as a transcriptional co-activator and an adaptor protein for intracellular adhesion. In epithelial tissues, β-catenin interacts with cadherins and α-catenin, and regulates epithelial cell growth and intracellular adhesion. By contrast, in Wnt signaling, β-catenin is a major transcriptional modulator and plays a crucial role in embryogenesis. In the absence of Wnt signaling, Apc forms a complex with β-catenin and Axin. This leads to phosphorylation of β-catenin by glycogen synthase kinase (GSK)-3β, which triggers degradation of β-catenin. When Wnt binds to its receptor Frizzled, Disheveled is hyper-phosphorylated, which results in release of

GSK3β from the β-catenin degradation complex and prevents phosphorylation of β-catenin by GSK3β. Unphosphorylated β-catenin translocates into the nucleus, where it forms a complex with Tcf/Lef and functions as a transcriptional co-activator.

In human cancers, such as colon cancer, the β-catenin/Tcf/Lef complex positively regulates the expression of a variety of cancer-associated genes, including *cyclin D, Tert,* and *c-Myc,* to promote tumorigenesis. Apc mutation leads to stabilization and accumulation of β-catenin in nuclei. However, in some cancers, mutation of β-catenin itself renders this protein unable to be phosphorylated, resulting in its stabilization.

Accumulated evidence suggests that Wnt/β-catenin signaling contributes to the maintenance of ES cell self-renewal. For example, *Apc*-null ES cells show severe differentiation defects [80]. Undifferentiated ES cells can be maintained in a self-renewing state by using conditioned medium from Wnt3a-expressing cells [81]. Furthermore, enforced expression of an activated form of β-catenin maintains the self-renewal of ES cells, even in the absence of LIF [23, 82]. Expression of γ-catenin, which has a similar structure to β-catenin, partially sustains the self-renewal of ES cells in the absence of LIF [23]. β-catenin binds to Oct3/4 to enhance its transcriptional activity in ES cells, leading to upregulation of *Nanog,* a target gene of Oct3/4 [23]. The Wnt/β-catenin pathway upregulates expression of *STAT3,* and this signaling converges with that of LIF [83]. Similarly, Wnt and LIF work in synergy to maintain the pluripotency of mouse ES cells [84]. On the other hand, the self-renewal of ES cells can be maintained without β-catenin [85]. Taken together, these results suggest that β-catenin promotes, but is dispensable, for ES cell self-renewal. Moreover, it was recently shown that ES cells lacking Wnt signaling resemble epiblast stem cells in terms of their morphology and gene expression [86, 87]. This suggests that Wnt/β-catenin signaling prevents the transition of ES cells from a naïve to a primed pluripotent state. In addition, β-catenin regulates Tert expression in ES cells [88], as is the case in cancer cells.

3.2. Gli

The zinc-finger transcription factor Gli is a central player in the Hedgehog (Hh)-mediated signaling pathway, which plays a critical role during embryogenesis. Gli belongs to the Klf family and has three isoforms in mammals, namely, Gli1, Gli2, and Gli3. Gli1 and Gli2 usually act as transcription activators, while Gli3 is a transcription suppressor. All Gli proteins have a DNA-binding domain consisting of C2-H2 class zinc fingers. In addition, Gli1 and Gli2 contain a C-terminal transactivation domain, while Gli2 and Gli3 have an N-terminal repression domain. Hh family proteins, namely, Sonic Hh, Indian Hh, and Desert Hh, function as ligands of the transmembrane receptor Ptch1. In the absence of a Hh ligand, Ptch1 inhibits the activity of the G-protein coupled receptor-like protein Smo, resulting in formation of a complex of Gli2 and Gli3 with the inhibitory protein Sufu. This results in cleavage of Gli2 and Gli3 into their repressor forms, which translocate into the nucleus. By contrast, binding of Hh ligands to Ptch1 results in the release and activation of Smo, leading to activation and nuclear translocation of Gli2, which results in transactivation of target genes, including *Gli1.*

Gli1 was originally identified as an amplified gene in a human glioma cell line [89]. It was recently shown that Hh/Gli signaling regulates the self-renewal of glioma stem cells, as well as their expression of stemness genes, including *Oct3/4* and *Sox2* [90]. Gli1 forms a positive feedback loop with Nanog, and the Nanog/Gli1 signaling axis is indispensable for regulation of glioma stem cells [91]. In Ewing sarcoma, expression of Gli1 is regulated in an Hh-independent manner: EWS-FLI, an oncogenic transcription factor that is produced by chromosomal translocation, directly upregulates *Gli1* expression to promote tumor growth [92, 93].

Gli1 and Gli2 are both highly expressed in undifferentiated ES cells, while Gli3 expression level is low [94]. Gli1 and Gli2 are downstream molecules of Oct3/4 and Nanog, and their expression is downregulated upon differentiation. When Gli1 and Gli2 are suppressed by a dominant-negative mutant of Gli2, expression of the self-renewal marker Sox2 decreases, whereas that of the differentiation markers Gata4 and Cdx2 increases, suggesting the importance of Gli activity for ES cell self-renewal. However, expression of this dominant-negative Gli2 mutant does not affect expression of the self-renewal markers Oct3/4 and Nanog. These findings suggest that Gli activity is involved in repressing ES cell differentiation, but is dispensable for ES cell self-renewal. In addition, Gli is involved in ES cell growth [94, 95].

3.3. Akt

The serine/threonine protein kinase Akt was independently identified by three different groups. Two groups identified this kinase as being homologous to protein kinase C and protein kinase A, giving rise to the names "protein kinase B" and "RAC-PK" (related to the A and C kinases) [96, 97]. The other group identified this kinase as the cellular counterpart of the oncogene *v-akt* of the acutely transforming retrovirus AKT8 that is found in a rodent T-cell lymphoma [98]. Here, I will describe this protein as "Akt". Akt has three isoforms, namely, Akt1, Akt2, and Akt3. Each Akt family member has an N-terminal pleckstrin homology domain, a short α-helical linker, and a C-terminal kinase domain. Akt is directly downstream of phosphatidylinositol-3-OH kinase, and is a key player in the regulation of cell growth and survival.

As expected from its identification as a counterpart of a viral oncogene, Akt plays an important role in human malignancy [99, 100]. Several studies have identified amplification of the *Akt* gene in human cancers. Amplification of *Akt1* was detected in a human gastric cancer [101]. Amplification and overexpression of Akt2 were detected in ovarian and pancreatic cancers. Artificial activation of Akt1 or Akt2 can transform NIH3T3 cells [102, 103], and Akt2 anti-sense RNA inhibits the tumorigenic phenotype of pancreatic carcinoma cell lines [104]. Furthermore, Akt1 kinase activity is often increased in prostate and breast cancers and is associated with a poor prognosis [102].

A constitutively activated Akt mutant can maintain the undifferentiated phenotype of mouse ES cells, even in the absence of LIF, although the mechanism underlying Akt-mediated maintenance of ES cell self-renewal is unclear [105]. Bechard and Dalton demonstrated that Akt phosphorylates, and thereby inactivates, GSK3β in ES cells, suggesting that Akt maintains ES cell self-renewal by inactivating GSK3β and thus stimulating activation of β-catenin [106]. By contrast, Watanabe *et al.* did not observe the accumulation of β-catenin in nuclei or activation of the transcriptional activity of β-catenin in mouse ES cells [105]. Another possible mechanism is that Akt induces expression of Tbx3, which in turn stimulates expression of Nanog to maintain ES cell self-renewal [107].

3.4. c-Myc

c-Myc is a cellular counterpart of the *v-myc* gene, which was isolated from the avian retrovirus MC29 [108], and belongs to a family of helix-loop-helix/leucine zipper transcription factors. c-Myc forms a complex with Max, which results in the increased stability of c-Myc protein [109]. Phosphorylation of Thr-62 by Erk also stabilizes c-Myc, while phosphorylation of Thr-58 by GSK3β reduces the stability of c-Myc [110]. c-Myc regulates expression of its target genes through binding to E-box sequences and recruiting histone acetyltransferases. Under normal conditions, when cells are stimulated by an internal or external growth-promoting signal, the level of c-Myc rapidly and transiently increases to induce cell proliferation, and the level of c-Myc subsequently returns to a low level in quiescent cells.

It is well-established that many, if not most, human tumors have elevated levels of c-Myc owing to gene amplification and translocation [111, 112]. Gene amplification of *c-Myc* has been reported in several cancers, including breast, ovarian, and colon. A common human translocation involving *c-Myc* is t(8;14), which is critical for the development of Burkitt's lymphoma. Although c-Myc was one of the four factors originally identified as being required for iPS cell production [19], generation of iPS cells without this transcription factor has been reported [113], suggesting that c-Myc is dispensable for cell reprogramming.

During the self-renewal of ES cells, levels of c-Myc are elevated [114]. By contrast, upon LIF withdrawal, the level of *c-Myc* mRNA decreases and c-Myc protein is phosphorylated on Thr-58 by GSK3β, which triggers degradation of c-Myc. Expression of a stable c-Myc mutant, in which the Thr-58 residue is mutated to alanine, allows ES cells to self-renew in the absence of LIF. By contrast, expression of a dominant-negative form of c-Myc inhibits the self-renewal of ES cells and induces their differentiation. These findings suggest that c-Myc is critically involved in maintaining the self-renewal of ES cells. Moreover, analysis of *Max*-null ES cells revealed that the function of c-Myc/Max in ES cell self-renewal seems to be largely independent of the Oct3/4, Sox2, and Nanog regulatory networks [115].

4. Future prospective

As I have described here, several common transcription factors are involved in the regulation of ES cell self-renewal and cancer cell growth (Fig. 2). This raises several intriguing possibilities. Considering that these common transcription factors are stem cell-specific and are involved in tumor growth, it is possible that they are specifically expressed in cancer stem cells and play important roles in the growth of these cells. Therefore, it is possible that these factors are good markers of cancer stem cells. If so, it might be possible to utilize these factors to isolate cancer stem cells, which will help to advance cancer stem cell research. Furthermore, identification of small compounds that specifically inhibit the functions of these factors may lead to the development of new anti-cancer drugs that can selectively kill cancer stem cells.

http://dx.doi.org/10.5772/57614

Since ES cells and iPS cells have similar gene expression profiles, it is likely that these common transcription factors are also expressed in iPS cells. Considering the transforming potential of these factors, it is possible that their high expression in iPS cells increases the risk of tumor formation during cell therapy using iPS-derived cells. Indeed, we have already found that *Zfp57*-null ES cells are significantly less able to form tumors than wild-type ES cells.

In this way, understanding the roles of putative oncogenes in ES cells will not only help to elucidate the molecular basis underlying the similarity between ES cells and cancers cells, but will also help to develop a novel method that can be used in cancer therapy and regenerative medicine.

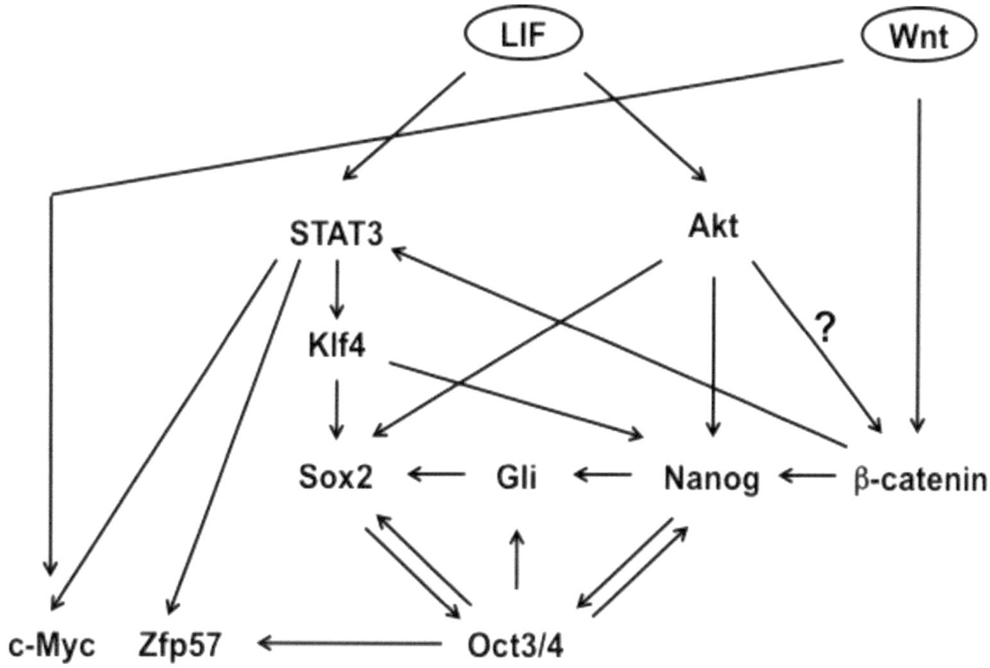

Figure 2. Transcription factor network that regulates ES cell self-renewal. Considering the similarities between ES cells and cancer cells, at least a part of this network may be used for growth regulation in cancer cells.

Author details

Hiroshi Koide*

Address all correspondence to: hkoide@med.kanazawa-u.ac.jp

Department of Stem Cell Biology, Graduate School of Medical Sciences, Kanazawa University, Kanazawa, Japan

References

[1] Evans MJ, Kauman MH. Establishment in culture of pluripotential cells from mouse embryos. Nature 1981; 292(5819) 154–156.

[2] Martin GR. Isolation of a pluripotent cell line from early mouse embryos cultured in medium conditioned by teratocarcinoma stem cells. Proceedings of the National Academy of Sciences USA 1981; 78(12) 7634–7638.

[3] Thomson JA, Itskovitz-Eldor J, Shapiro SS, Waknitz MA, Swiergiel JJ, Marshall VS, Jones JM. Embryonic stem cell lines derived from human blastocysts. Science 1998; 282(5391) 1145-1147.

[4] Takahashi K, Tanabe K, Ohnuki M, Narita M, Ichisaka T, Tomoda K, Yamanaka S. Induction of pluripotent stem cells from adult human fibroblasts by defined factors. Cell. 2007; 131(5) 861-872.

[5] Yu J, Vodyanik MA, Smuga-Otto K, Antosiewicz-Bourget J, Frane JL, Tian S, Nie J, Jonsdottir GA, Ruotti V, Stewart R, Slukvin II, Thomson JA. Induced pluripotent stem cell lines derived from human somatic cells. Science 2007; 318(5858) 1917-1920.

[6] Smith AG, Heath JK, Donaldson DD, Wong GG, Moreau J, Stahl M, Rogers D. Inhibition of pluripotential embryonic stem cell differentiation by purified polypeptides. Nature 1988; 336(6200) 688-690.

[7] Williams RL, Hilton DJ, Pease S, Willson TA, Stewart CL, Gearing DP, Wagner EF, Metcalf D, Nicola NA, Gough NM. Myeloid leukaemia inhibitory factor maintains the developmental potential of embryonic stem cells. Nature 1988; 336(6200) 684-687.

[8] Koide H, Yokota T. The LIF/STAT3 pathway in ES cell self-renewal. In: Atwood CS. (ed) Embryonic stem cells-The hormonal regulation of pluripotency and embryogenesis. Rijeka: InTech; 2011. p61-78.

[9] Niwa H. How is pluripotency determined and maintained? Development 2007; 134(4) 635-646.

[10] Lapidot T, Sirard C, Vormoor J, Murdoch B, Hoang T, Caceres-Cortes J, Minden M, Paterson B, Caligiuri MA, Dick JE. A cell initiating human acute myeloid leukaemia after transplantation into SCID mice. Nature 1994; 367(6464) 645-648.

[11] Al-Hajj M, Wicha MS, Benito-Hernandez A, Morrison SJ, Clarke MF. Proceedings of the National Academy of Sciences USA 2003; 100(7) 3983–3988.

[12] Singh SK, Clarke ID, Terasaki M, Bonn VE, Hawkins C, Squire J, Dirks PB. Identification of a cancer stem cell in human brain tumors. Cancer Research 2003; 63(18) 5821-5828.

[13] Ben-Porath I, Thomson MW, Carey VJ, Ge R, Bell GW, Regev A, Weinberg RA. An embryonic stem cell–like gene expression signature in poorly differentiated aggressive human tumors. Nature Genetics 2008; 40(5) 499-507.

[14] Schöler HR, Hatzopoulos AK, Balling R, Suzuki N, Gruss P. A family of octamer-specific proteins present during mouse embryogenesis: evidence for germline-specific expression of an Oct factor. EMBO Journal 1989; 8(9) 2543-2550.

[15] Okamoto K, Okazawa H, Okuda A, Sakai M, Muramatsu M, Hamada H. A novel octamer binding transcription factor is differentially expressed in mouse embryonic cells. Cell 1990; 60(3) 461-472.

[16] Rosner MH, Vigano MA, Ozato K, Timmons PM, Poirier F, Rigby PW, Staudt LM. A POU-domain transcription factor in early stem cells and germ cells of the mammalian embryo. Nature 1990; 345(6277) 686-692.

[17] Nichols J, Zevnik B, Anastassiadis K, Niwa H, Klewe-Nebenius D, Chambers I, Schöler H, Smith A. Formation of pluripotent stem cells in the mammalian embryo depends on the POU transcription factor Oct4. Cell. 1998; 95(3) 379-391.

[18] Niwa H, Miyazaki J, Smith AG. Quantitative expression of Oct-3/4 defines differentiation, dedifferentiation or self-renewal of ES cells. Nature Genetics 2000; 24(4) 372–376.

[19] Takahashi K, Yamanaka S. Induction of pluripotent stem cells from mouse embryonic and adult fibroblast cultures by defined factors. Cell 2006; 126(4) 663-676.

[20] Yeom YI, Fuhrmann G, Ovitt CE, Brehm A, Ohbo K, Gross M, Hubner K, Scholer HR. Germline regulatory element of Oct-4 specific for the totipotent cycle of embryonal cells. Development 1996; 122(3) 881–894.

[21] Gu P, Goodwin B, Chung AC, Xu X, Wheeler DA, Price RR, Galardi C, Peng L, Latour AM, Koller BH, Gossen J, Kliewer SA, Cooney AJ. Orphan nuclear receptor LRH-1 is required to maintain Oct4 expression at the epiblast stage of embryonic development. Molecular and Cellular Biology 2005; 25(9) 3492–3505.

[22] Okumura-Nakanishi S, Saito M, Niwa H, Ishikawa F. Oct-3/4 and Sox2 regulate Oct-3/4 gene in embryonic stem cells. Journal of Biological Chemistry 2005; 280(7) 5307–5317.

[23] Takao Y, Yokota T, Koide H. Beta-catenin up-regulates Nanog expression through interaction with Oct-3/4 in embryonic stem cells. Biochemical and Biophysical Research Communications 2007; 353(3) 699-705.

[24] Sun C, Nakatake Y, Akagi T, Ura H, Matsuda T, Nishiyama A, Koide H, Ko MS, Niwa H, Yokota T. Dax1 binds to Oct3/4 and inhibits its transcriptional activity in embryonic stem cells. Molecular and Cellular Biology 2009; 29(16) 4574-4583.

[25] Liu A, Yu X, Liu S. Pluripotency transcription factors and cancer stem cells: small genes make a big difference. Chinese Journal of Cancer 2013; 32(9) 483-487.

[26] Gibbs CP, Kukekov VG, Reith JD, Tchigrinova O, Suslov ON, Scott EW, Ghivizzani SC, Ignatova TN, Steindler DA. Stem-like cells in bone sarcomas: Implications for tumorigenesis. Neoplasia 2005; 7(11) 967–976.

[27] Ponti D, Costa A, Zaffaroni N, Pratesi G, Petrangolini G, Coradini D, Pilotti S, Pierotti MA, Daidone MG. Isolation and in vitro propagation of tumorigenic breast cancer cells with stem/progenitor cell properties. Cancer Research 2005; 65(13) 5506–5511.

[28] Chen YC, Hsu HS, Chen YW, Tsai TH, How CK, Wang CY, Hung SC, Chang YL, Tsai ML, Lee YY, Ku HH, Chiou SH. Oct-4 expression maintained cancer stem-like properties in lung cancer-derived CD133-positive cells. PLoS One 2008; 3(7) e2637.

[29] Luo W, Li S, Peng B, Ye Y, Deng X, Yao K. Embryonic stem cells markers SOX2, OCT4 and Nanog expression and their correlations with epithelial-mesenchymal transition in nasopharyngeal carcinoma. PLoS One 2013; 8(2) e56324.

[30] Jeon HM, Sohn YW, Oh SY, Kim SH, Beck S, Kim S, Kim H. ID4 imparts chemoresistance and cancer stemness to glioma cells by derepressing miR-9*-mediated suppression of SOX2. Cancer Research 2011; 71(9) 3410-3421.

[31] Tsuruzoe S, Ishihara K, Uchimura Y, Watanabe S, Sekita Y, Aoto T, Saitoh H, Yuasa Y, Niwa H, Kawasuji M, Baba H, Nakao M. Inhibition of DNA binding of Sox2 by the SUMO conjugation. Biochemical and Biophysical Research Communications 2006; 351 (4) 920–926.

[32] Baltus GA, Kowalski MP, Zhai H, Tutter AV, Quinn D, Wall D, Kadam S. Acetylation of sox2 induces its nuclear export in embryonic stem cells. Stem Cells 2009;27(9) 2175-2184.

[33] Jeong CH, Cho YY, Kim MO, Kim SH, Cho EJ, Lee SY, Jeon YJ, Lee KY, Yao K, Keum YS, Bode AM, Dong Z. Phosphorylation of Sox2 cooperates in reprogramming to pluripotent stem cells. Stem Cells 2010; 28(12) 2141-2150.

[34] Avilion AA, Nicolis SK, Pevny LH, Perez L, Vivian N, Lovell-Badge R. Multipotent cell lineages in early mouse development depend on SOX2 function. Genes and Development 2003; 17(1) 126-140.

[35] Masui S, Nakatake Y, Toyooka Y, Shimosato D, Yagi R, Takahashi K, Okochi H, Okuda A, Matoba R, Sharov AA, Ko MSH, Niwa H. Pluripotency governed by Sox2 via regulation of Oct3/4 expression in mouse embryonic stem cells. Nature Cell Biology 2007; 9(6) 625-635.

[36] Ura H, Murakami K, Akagi T, Kinoshita K, Yamaguchi S, Masui S, Niwa H, Koide H, Yokota T. Eed/Sox2 regulatory loop controls ES cell self-renewal through histone methylation and acetylation. EMBO Journal 2011 ; 30(11) 2190-2204.

[37] Liu K, Lin B, Zhao M, Yang X, Chen M, Gao A, Liu F, Que J, Lan X. The multiple roles for Sox2 in stem cell maintenance and tumorigenesis. Cell Signal 2013; 25(5) 1264-1271.

[38] Bass AJ, Watanabe H, Mermel CH, Yu S, Perner S, Verhaak RG, Kim SY, Wardwell L, Tamayo P, Gat-Viks I, Ramos AH, Woo MS, Weir BA, Getz G, Beroukhim R, O'Kelly M, Dutt A, Rozenblatt-Rosen O, Dziunycz P, Komisarof J, Chirieac LR, Lafargue CJ, Scheble V, Wilbertz T, Ma C, Rao S, Nakagawa H, Stairs DB, Lin L, Giordano TJ, Wagner P, Minna JD, Gazdar AF, Zhu CQ, Brose MS, Cecconello I, Jr UR, Marie SK, Dahl O, Shivdasani RA, Tsao MS, Rubin MA, Wong KK, Regev A, Hahn WC, Beer DG, Rustgi AK, Meyerson M. SOX2 is an amplified lineage-survival oncogene in lung and esophageal squamous cell carcinomas. Nature Genetics 2009; 41(11) 1238-1242.

[39] Lengerke C, Fehm T, Kurth R, Neubauer H, Scheble V, Müller F, Schneider F, Petersen K, Wallwiener D, Kanz L, Fend F, Perner S, Bareiss PM, Staebler A. Expression of the embryonic stem cell marker SOX2 in early-stage breast carcinoma. BMC Cancer 2011; 11 42.

[40] Leis O, Eguiara A, Lopez-Arribillaga E, Alberdi MJ, Hernandez-Garcia S, Elorriaga K, Pandiella A, Rezola R, Martin AG. Sox2 expression in breast tumours and activation in breast cancer stem cells. Oncogene 2012; 31(11) 1354-1365.

[41] Bareiss PM, Paczulla A, Wang H, Schairer R, Wiehr S, Kohlhofer U, Rothfuss OC, Fischer A, Perner S, Staebler A, Wallwiener D, Fend F, Fehm T, Pichler B, Kanz L, Quintanilla-Martinez L, Schulze-Osthoff K, Essmann F, Lengerke C. SOX2 expression associates with stem cell state in human ovarian carcinoma. Cancer Research 2013; 73(17) 5544-5555.

[42] Chambers I, Colby D, Robertson M, Nichols J, Lee S, Tweedie S, Smith A. Functional expression cloning of Nanog, a pluripotency sustaining factor in embryonic stem cells. Cell 2003; 113(5) 643-655.

[43] Mitsui K, Tokuzawa Y, Itoh H, Segawa K, Murakami M, Takahashi K, Maruyama M, Maeda M, Yamanaka S. The homeoprotein Nanog is required for maintenance of pluripotency in mouse epiblast and ES cells. Cell 2003; 113(5) 631-642.

[44] Ramakrishna S, Suresh B, Lim KH, Cha BH, Lee SH, Kim KS, Baek KH. PEST motif sequence regulating human NANOG for proteasomal degradation. Stem Cells and Development 2012; 20(9) 1511–1519.

[45] Moretto-Zita M, Jin H, Shen Z, Zhao T, Briggs SP, Xu Y. Phosphorylation stabilizes Nanog by promoting its interaction with Pin1. Proceedings of the National Academy of Sciences USA 2010; 107(30) 13312–13317.

[46] Chambers I, Silva J, Colby D, Nichols J, Nijmeijer B, Robertson M, Vrana J, Jones K, Grotewold L, Smith A. Nanog safeguards pluripotency and mediates germline development. Nature 2007; 450(7173) 1230-1234.

[47] Silva J, Nichols J, Theunissen TW, Guo G, van Oosten AL, Barrandon O, Wray J, Yamanaka S, Chambers I, Smith A. Nanog is the gateway to the pluripotent ground state. Cell 2009; 138(4) 722–737.

[48] Wang ML, Chiou SH, Wu CW. Targeting cancer stem cells: emerging role of Nanog transcription factor. Onco Targets and Therapy 2013; 6 1207-1220.

[49] Hart AH, Hartley L, Parker K, Ibrahim M, Looijenga LH, Pauchnik M, Chow CW, Robb L. The pluripotency homeobox gene NANOG is expressed in human germ cell tumors. Cancer 2005; 104(10) 2092-2098.

[50] Siu MK, Wong ES, Kong DS, Chan HY, Jiang L, Wong OG, Lam EW, Chan KK, Ngan HY, Le XF, Cheung AN. Stem cell transcription factor NANOG controls cell migration and invasion via dysregulation of E-cadherin and FoxJ1 and contributes to adverse clinical outcome in ovarian cancers. Oncogene 2013; 32(30) 3500-3509.

[51] Sun C, Sun L, Jiang K, Gao DM, Kang XN, Wang C, Zhang S, Huang S, Qin X, Li Y, Liu YK. NANOG promotes liver cancer cell invasion by inducing epithelial-mesenchymal transition through NODAL/SMAD3 signaling pathway. International Journal of Biochemistry and Cell Biology 2013; 45(6) 1099-1108.

[52] Zhang J, Wang X, Li M, Han J, Chen B, Wang B, Dai J. NANOGP8 is a retrogene expressed in cancers. FEBS Journal 2006; 273(8) 1723-1730.

[53] Jeter CR, Badeaux M, Choy G, Chandra D, Patrawala L, Liu C, Calhoun-Davis T, Zaehres H, Daley GQ, Tang DG. Functional evidence that the self-renewal gene NANOG regulates human tumor development. Stem Cells 2009; 27(5) 993-1005.

[54] Palla AR, Piazzolla D, Abad M, Li H, Dominguez O, Schonthaler HB, Wagner EF, Serrano M. Reprogramming activity of NANOGP8, a NANOG family member widely expressed in cancer. Oncogene 2013; doi: 10.1038/onc.2013.196.

[55] Wegenka UM, Buschmann J, Lutticken C, Heinrich PC, Horn F. Acute-phase response factor, a nuclear factor binding to acute-phase response elements, is rapidly activated by interleukin-6 at the posttranslational level. Molecular and Cellular Biology; 13(1) 276-288.

[56] Duncan SA, Zhong Z, Wen Z, Darnell Jr JE. STAT signaling is active during early mammalian development. Developmental Dynamics 1997; 208(2) 190-198.

[57] Takeda K, Noguchi K, Shi W, Tanaka T, Matsumoto M, Yoshida N, Kishimoto T, Akira S. Targeted disruption of the mouse Stat3 gene leads to early embryonic lethality. Proceedings of the National Academy of Sciences USA 1997; 94(8) 3801-3804.

[58] Yu H, Jove R. The STATs of cancer--new molecular targets come of age. Nature Reviews Cancer 2004; 4(2) 97-105.

[59] Shields JM, Christy RJ, Yang VW. Identification and characterization of a gene encoding a gut-enriched Krüppel-like factor expressed during growth arrest. Journal of Biological Chemistry 1996; 271(33) 20009–20017.

[60] Garrett-Sinha LA, Eberspaecher H, Seldin MF, de Crombrugghe B. A gene for a novel zinc-finger protein expressed in differentiated epithelial cells and transiently in certain mesenchymal cells. Journal of Biological Chemistry 1996; 271(49) 31384–31390.

[61] Nakatake Y, Fukui N, Iwamatsu Y, Masui S, Takahashi K, Yagi R, Yagi K, Miyazaki J, Matoba R, Ko MS, Niwa H. KLF4 cooperates with Oct3/4 and Sox2 to activate the Lefty1 core promoter in embryonic stem cells. Molecular and Cellular Biology 2006; 26(20) 7772–7782.

[62] Jiang J, Chan YS, Loh YH, Cai J, Tong GQ, Lim CA, Robson P, Zhong S, Ng HH. A core KLF circuitry regulates self-renewal of embryonic stem cells. Nature Cell Biology 2008; 10(3) 353–360.

[63] Li Y, McClintick J, Zhong L, Edenberg HJ, Yoder MC, Chan RJ. Murine embryonic stem cell differentiation is promoted by SOCS-3 and inhibited by the zinc finger transcription factor KLF4. Blood 2005; 105(2) 635–637.

[64] Zhang P, Andrianakos R, Yang Y, Liu C, Lu W. Kruppel-like factor 4 prevents embryonic stem cell differentiation by regulating Nanog gene expression. Journal of Biological Chemistry 2010; 285(12) 9180-9189.

[65] Segre JA, Bauer C, Fuchs E. KLF4 is a transcription factor required for establishing the barrier function of the skin. Nature Genetics 1999; 22(4) 356–260.

[66] Zhao W, Hisamuddin IM, Nandan MO, Babbin BA, Lamb NE, Yang VW. Identification of Krüppel-like factor 4 as a potential tumor suppressor gene in colorectal cancer. Oncogene 2004; 23(2) 395–402.

[67] Wei D, Gong W, Kanai M, Schlunk C, Wang L, Yao JC, Wu TT, Huang S, Xie K. Drastic down-regulation of Krüppel-like factor 4 expression is critical in human gastric cancer development and progression. Cancer Research 2005; 65(7) 2746–2754.

[68] Foster KW, Liu Z, Nail CD, Li X, Fitzgerald TJ, Bailey SK, Frost AR, Louro ID, Townes TM, Paterson AJ, Kudlow JE, Lobo-Ruppert SM, Ruppert JM. Induction of KLF4 in basal keratinocytes blocks the proliferation-differentiation switch and initiates squamous epithelial dysplasia. Oncogene 2005; 24(9) 1491–1500.

[69] Huang CC, Liu Z, Li X, Bailey SK, Nail CD, Foster KW, Frost AR, Ruppert JM, Lobo-Ruppert SM. KLF4 and PCNA identify stages of tumor initiation in a conditional model of cutaneous squamous epithelial neoplasia. Cancer Biology and Therapy 2005; 4(12) 1401–1408.

[70] Yin X, Li YW, Jin JJ, Zhou Y, Ren ZG, Qiu SJ, Zhang BH. The clinical and prognostic implications of pluripotent stem cell gene expression in hepatocellular carcinoma. Oncology Letters 2013; 5(4) 1155-1162.

[71] Okazaki S, Tanase S, Choudhury BK, Setoyama K, Miura R, Ogawa M, Setoyama C. A novel nuclear protein with zinc fingers down-regulated during early mammalian cell differentiation. Journal of Biological Chemistry 1994; 269(9) 6900-6907.

[72] Alonso MB, Zoidl G, Taveggia C, Bosse F, Zoidl C, Rahman M, Parmantier E, Dean CH, Harris BS, Wrabetz L, Muller HW, Jessen KR, Mirsky R. Identification and characterization of ZFP-57, a novel zinc finger transcription factor in the mammalian peripheral nervous system. Journal of Biological Chemistry 2004; 279(24) 25653-25664.

[73] Li X, Ito M, Zhou F, Youngson N, Zuo X, Leder P, Ferguson-Smith AC. A maternal-zygotic effect gene, Zfp57, maintains both maternal and paternal imprints. Developmental Cell 2008; 15(4) 547-557.

[74] Quenneville S, Verde G, Corsinotti A, Kapopoulou A, Jakobsson J, Offner S, Baglivo I, Pedone PV, Grimaldi G, Riccio A, Trono D. In embryonic stem cells, ZFP57/KAP1 recognize a methylated hexanucleotide to affect chromatin and DNA methylation of imprinting control regions. Molecular Cell 2011; 44(3) 361-372.

[75] Mackay DJ, Callaway JL, Marks SM, White HE, Acerini CL, Boonen SE, Dayanikli P, Firth HV, Goodship JA, Haemers AP, Hahnemann JM, Kordonouri O, Masoud AF, Oestergaard E, Storr J, Ellard S, Hattersley AT, Robinson DO, Temple IK. Hypomethylation of multiple imprinted loci in individuals with transient neonatal diabetes is associated with mutations in ZFP57. Nature Genetics 2008; 40(8) 949-951.

[76] Akagi T, Usuda M, Matsuda T, Ko MS, Niwa H, Asano M, Koide H, Yokota T. Identification of Zfp-57 as a downstream molecule of STAT3 and Oct-3/4 in embryonic stem cells. Biochemical and Biophysical Research Communications 2005; 331(1) 23-30.

[77] Tada Y, Yamaguchi Y, Kinjo T, Song X, Akagi T, Takamura H, Ohta T, Yokota T, Koide H. The stem cell transcription factor ZFP57 induces IGF2 expression to promote anchorage-independent growth in cancer cells. Oncogene 2014 (*in press*).

[78] Kinzler KW, Nilbert MC, Su LK, Vogelstein B, Bryan TM, Levy DB, Smith KJ, Preisinger AC, Hedge P, Mckechnie D, Finniear R, Markham A, Groffen J, Boguski MS, Altschul SF, Horii A, Ando H, Miyoshi Y, Mild Y, Nishisho I, Nakamura Y. Identification of FAP locus genes from Chromosome 5q21. Science 1991; 253(5020) 661-665.

[79] Nishisho I, Nakamura Y, Miyoshi Y, Miki Y, Ando H, Horii A, Koyama K, Utsunomiya J, Baba S, Hedge P, Markham A, Krush AJ, Pctersen G, Hamilton SR, Nilbert MC, Levy DB, Bryan TM, Preisinger AC, Smith KJ, Su LK, Kinzler KW, Vogelstein B. Mutations of chromosome 5q21 genes in FAP and colorectal cancer patients. Science 1991; 253(5020) 665-669.

[80] Kielman MF, Rindapää M, Gaspar C, van Poppel N, Breukel C, van Leeuwen S, Taketo MM, Roberts S, Smits R, Fodde R. Apc modulates embryonic stem-cell differentiation by controlling the dosage of beta-catenin signaling. Nature Genetics 2002; 32(4) 594-605.

[81] Singla DK, Schneider DJ, LeWinter MM, Sobel BE. Wnt3a but not wnt11 supports self-renewal of embryonic stem cells. Biochemical and Biophysical Research Communications 2006; 345(2) 789-795.

[82] Kelly KF, Ng DY, Jayakumaran G, Wood GA, Koide H, Doble BW. β-catenin enhances Oct-4 activity and reinforces pluripotency through a TCF-independent mechanism. Cell Stem Cell 2011; 8(2) 214-227.

[83] Hao J, Li TG, Qi X, Zhao DF, Zhao GO. WNT/beta-catenin pathway up-regulates Stat3 and converges on LIF to prevent differentiation of mouse embryonic stem cells. Developmental Biology 2006; 290(1) 81-91.

[84] Ogawa K, Nishinakamura R, Iwamatsu Y, Shimosato D, Niwa H. Synergistic action of Wnt and LIF in maintaining pluripotency of mouse ES cells. Biochemical and Biophysical Research Communications 2006; 343(1) 159-166.

[85] Lyashenko N, Winter M, Migliorini D, Biechele T, Moon RT, Hartmann C. Differential requirement for the dual functions of β-catenin in embryonic stem cell self-renewal and germ layer formation. Nature Cell Biology 2011; 13(7) 753-761.

[86] Anton R, Kestler HA, Kuhl M. Beta-catenin signaling contributes to stemness and regulates early differentiation in murine embryonic stem cells. FEBS Letters 2007; 581(27) 5247–5254.

[87] ten Berge D, Kurek D, Blauwkamp T, Koole W, Maas A, Eroglu E, Siu RK, Nusse R. Embryonic stem cells require Wnt proteins to prevent differentiation to epiblast stem cells. Nature Cell Biology 2011; 13(9) 1070-1075.

[88] Hoffmeyer K, Raggioli A, Rudloff S, Anton R, Hierholzer A, Del Valle I, Hein K, Vogt R, Kemler R. Wnt/β-catenin signaling regulates telomerase in stem cells and cancer cells. Science 2012; 336(6088) 1549-1554.

[89] Kinzler KW, Bigner SH, Bigner DD, Trent JM, Law ML, O'Brien SJ, Wong AJ, Vogelstein B. Identification of an amplified, highly expressed gene in a human glioma. Science 1987; 236(4797) 70–73.

[90] Clement V, Sanchez P, de Tribolet N, Radovanovic I, Ruiz i Altaba A. HEDGEHOG-GLI1 signaling regulates human glioma growth, cancer stem cell self-renewal, and tumorigenicity. Current Biology 2007; 17(2) 165-172.

[91] Zbinden M, Duquet A, Lorente-Trigos A, Ngwabyt SN, Borges I, Ruiz i Altaba A. NANOG regulates glioma stem cells and is essential in vivo acting in a crossfunctional network with GLI1 and p53. EMBO Journal 2010; 29(15) 2659–2674.

[92] Zwerner JP, Joo J, Warner KL, Christensen L, Hu-Lieskovan S, Triche TJ, May WA. The EWS/FLI1 oncogenic transcription factor deregulates GLI1. Oncogene 2008; 27(23) 3282-3291.

[93] Beauchamp E, Bulut G, Abaan O, Chen K, Merchant A, Matsui W, Endo Y, Rubin JS, Toretsky J, Uren A. GLI1 is a direct transcriptional target of EWS-FLI1 oncoprotein. Journal of Biological Chemistry 2009; 284(14) 9074–9082.

[94] Ueda A. Involvement of Gli proteins in undifferentiated state maintenance and proliferation of embryonic stem cells. Journal of the Juzen Medical Society 2012; 121(2) 38-46.

[95] Heo JS, Lee MY, Han HJ. Sonic hedgehog stimulates mouse embryonic stem cell proliferation by cooperation of Ca^{2+}/Protein kinase C and epidermal growth factor receptor as well as Gli1 activation. Stem Cells 2007; 25(12) 3069-3080.

[96] Coffer PJ, Woodgett JR: Molecular cloning and charecterisation of a novel protein-serine kinase related to the cAMPdependent and protein kinase C families. European Journal of Biochemistry 1991; 201(2) 475-481.

[97] Jones PF, Jakubowicz T, Pitossi FJ, Maurer F, Hemmings BA. Molecular cloning and identification of a serine/threonine protein kinase of the second-messenger subfamily. Proceedings of the National Academy of Sciences USA 1991; 88(10) 4171-4175.

[98] Bellacosa A, Testa JR, Staal SP, Tsichlis PN. A retroviral oncogene, akt, encoding a serine-threonine kinase containing an SH2-1ike region. Science 1991; 254(5029) 274-277.

[99] Testa JR, Bellacosa A. AKT plays a central role in tumorigenesis. Proceedings of the National Academy of Sciences USA 2001; 98(20) 10983-10985.

[100] Nicholson KM, Anderson NG. The protein kinase B/Akt signalling pathway in human malignancy. Cellular Signalling 2002; 14(5) 381-395.

[101] Staal SP. Molecular cloning of the akt oncogene and its human homologues AKT1 and AKT2: amplification of AKT1 in a primary human gastric adenocarcinoma. Proceedings of the National Academy of Sciences USA 1987; 84(14) 5034–5037.

[102] Sun M, Wang G, Paciga JE, Feldman RI, Yuan ZQ, Ma XL, Shelley SA, Jove R, Tsichlis PN, Nicosia SV, Cheng JQ. AKT1/PKBalpha kinase is frequently elevated in human cancers and its constitutive activation is required for oncogenic transformation in NIH3T3 cells. American Journal of Pathology 2001; 159(2) 431–437.

[103] Cheng JQ, Altomare DA, Klein MA, Lee WC, Kruh GD, Lissy NA, Testa JR. Transforming activity and mitosis-related expression of the AKT2 oncogene: evidence suggesting a link between cell cycle regulation and oncogenesis. Oncogene 1997; 14(23) 2793–2801.

[104] Cheng JQ, Ruggeri B, Klein WM, Sonoda G, Altomare DA, Watson DK, Testa JR. Amplification of AKT2 in human pancreatic cells and inhibition of AKT2 expression

and tumorigenicity by antisense RNA. Proceedings of the National Academy of Sciences USA 1996; 93(8) 3636-3641.

[105] Watanabe S, Umehara H, Murayama K, Okabe M, Kimura T, Nakano T. Activation of Akt signaling is sufficient to maintain pluripotency in mouse and primate embryonic stem cells. Oncogene 2006; 25(19) 2697-2707.

[106] Bechard M, Dalton S. Subcellular localization of glycogen synthase kinase 3beta controls embryonic stem cell self-renewal. Molecular and Cellular Biology 2009; 29(8) 2092-2104.

[107] Niwa H, Ogawa K, Shimosato D, Adachi K. A parallel circuit of LIF signalling pathways maintains pluripotency of mouse ES cells. Nature 2009; 460(7251) 118-122.

[108] Alitalo K, Bishop JM, Smith DH, Chen EY, Colby WW, Levinson AD. Nucleotide sequence to the v-myc oncogene of avian retrovirus MC29. Proceedings of the National Academy of Sciences USA 1983; 80(1) 100-104.

[109] Blackwood EM, Luscher B, Eisenman RN. Myc and Max associate in vivo. Genes and Development 1992; 6(1) 71-80.

[110] Sears, R., Nuckolls, F., Haura, E., Taya, Y., Tamai, K., and Nevins, J.R. (2000). Multiple Ras-dependent phosphorylation pathways regulate Myc protein stability. Genes and Development 14, 2501–2514.

[111] Alitalo K, Koskinen P, Mäkelä TP, Saksela K, Sistonen L, Winqvist R. myc oncogenes: activation and amplification. Biochimica et Biophysica Acta 1987; 907(1) 1-32.

[112] Brison O. Gene amplification and tumor progression. Biochimica et Biophysica Acta 1993; 1155(1) 25-41.

[113] Nakagawa M, Koyanagi M, Tanabe K, Takahashi K, Ichisaka T, Aoi T, Okita K, Mochiduki Y, Takizawa N, Yamanaka S. Generation of induced pluripotent stem cells without Myc from mouse and human fibroblasts. Nature Biotechnology 2008; 26(1) 101-1016.

[114] Cartwright P, McLean C, Sheppard A, Rivett D, Jones K, Dalton S. LIF/STAT3 controls ES cell self-renewal and pluripotency by a Myc-dependent mechanism. Development 2005; 132(5) 885-896.

[115] Hishida T, Nozaki Y, Nakachi Y, Mizuno Y, Okazaki Y, Ema M, Takahashi S, Nishimoto M, Okuda A. Indefinite self-renewal of ESCs through Myc/Max transcriptional complex-independent mechanisms. Cell Stem Cell. 2011; 9(1) 37-49.

Section 2

Stem Cells as a Model System for Studing Haematopoiesis

Chapter 3

Embryonic Stem Cell Differentiation – A Model System to Study Embryonic Haematopoiesis

Monika Stefanska, Valerie Kouskoff and Georges Lacaud

Additional information is available at the end of the chapter

http://dx.doi.org/10.5772/57607

1. Introduction

Haematopoiesis, the process of generation of blood cells, is one of the most extensively studied developmental systems. The whole spectrum of blood cells produced in mammalian organisms includes primitive erythrocytes and definitive haematopoietic cells such as myeloid, lymphoid, definitive erythroid and haematopoietic stem cells (HSCs).

Haematopoiesis takes place in several locations during ontogeny and in adult life. The embryonic origin of blood cells has been studied for more than a century. However, studies on haematopoiesis *in vivo* are challenging as embryos, and in particular mammalian embryos, are extremely small and difficult to access at these very early stages of development. Moreover, the number of cells per embryo is limited and all the successive developmental events take place very fast. Therefore different approaches have been developed to facilitate these studies *in vitro* and one of them involves the use of embryonic stem (ES) cell *in vitro* differentiation. In this chapter, we will highlight some recent results on studies of the development of the haematopoietic system obtained in particular using the *in vitro* differentiation of murine ES cells. We will also present the methods we routinely use in our laboratory to work with wild type or genetically modified murine ES cells.

2. Early haematopoietic development

2.1. How can embryonic stem cells be used to study early embryonic haematopoiesis?

Studying haematopoiesis in the mouse embryo *in vivo* remains challenging, in particular at the very early stages of development when the embryo is small, difficult to access and the number of cells is limited. One of the alternative approaches is the *in vitro* differentiation of ES cells which are defined as pluripotent cells, able to give rise to three primary germ cell layers (endoderm, mesoderm and ectoderm) [1]. ES cells are isolated from the inner cell mass of the blastocyst and under appropriate conditions can be maintained undifferentiated in culture [2] or alternatively allowed to differentiate. By scaling up cultures of differentiated ES cells, it is relatively easy to access large number of cells that would be unattainable *in vivo*. ES cells represent a unique tool to study the molecular and cellular mechanisms of normal haematopoietic development, or the perturbations of these mechanisms leading to pathogenesis. In addition, with the advent of human ES cells, and induced human pluripotent stem (iPS) cells, the differentiation of these stem cells toward haematopoiesis could represent an exciting approach to generate cell populations to treat haematological disorders.

2.2. Sites of haematopoietic development

In 1920, the embryologist Florence Sabin observed that endothelial and haematopoietic cells were closely located in the yolk sac of avian embryo [3]. These structures, later called "blood islands", were thought to be derived from mesodermal cells undergoing differentiation towards endothelial and haematopoietic lineages [4]. In the mouse embryo, the first blood cells were shown to emerge around day E7.5 in the extra-embryonic yolk sac, within the blood islands [5]. These first haematopoietic cells are primitive erythrocytes that transport large amounts of oxygen required to support the rapidly growing embryo. In the final days of gestation, their number decreases rapidly as other haematopoietic cells overtake their function.

For a long time, the yolk sac was thought to generate only primitive erythrocytes. However, detailed studies indicated that other cell lineages such as definitive erythroid progenitors, mast cells and bipotential granulocyte/macrophage progenitors are also generated in the murine yolk sac before circulation [6]. By day E8.5, circulation in the mouse embryo is established and the newly formed blood vessels connect the extra-embryonic yolk sac to intra-embryonic tissues. From that time onward, other haematopoietic tissues within the embryo proper become actively involved in haematopoiesis. In 1994, a seminal study by Muller and co-workers demonstrated that at day 10 p.c. (post coitus), the aorta-gonad-mesonephros (AGM) region contains long term repopulating haematopoietic stem cells (HSCs)-the foundation of the blood system in adult organisms [7]. The AGM region is an intra-embryonic site that will later develop into major internal organs. Following the discovery of these first HSCs in the AGM region it was important to distinguish whether these cells were generated in this region or emigrated from other embryonic locations through the blood circulation. The work of Medvinsky and Dzierzak established that definitive HSCs, capable of long term multilineage haematopoietic reconstitution emerge but also expand within the AGM region [8]. More recently, the placenta, both in mouse and human was reported to contain HSCs [9, 10], although whether these HSCs are *de novo* generated within the placenta remains unknown. During adult life, the main site of haematopoiesis is the bone marrow, where HSCs are found. The capacity of HSCs present in the bone marrow to rebuild the whole haematopoietic hierarchy in recipient organism is routinely used by clinicians to treat many blood-related diseases through bone marrow transplantations.

2.3. Haemangioblast and haemogenic endothelium – Is there a connection?

The search for the cellular origin of blood cells started nearly 100 years ago, when Sabin noticed that endothelial and haematopoietic lineages are located in close proximity within the blood islands, suggesting the existence of a common precursor called a haemangioblast [3]. Few years later, in 1924 Alexander Maximow observed that the blood islands represent mesodermal masses that differentiated towards endothelial and haematopoietic cells [4]. *In vitro* experiments based on embryonic stem (ES) cell differentiation were the first experiments providing substantial data supporting the existence of the haemangioblast [11-13]. First, Choi and co-workers identified a precursor called blast colony forming cell (BL-CFC), expressing FLK1 – the VEGR receptor 2, that upon culture gave rise to blast colonies containing precursors for both endothelial and haematopoietic cells [11]. These BL-CFCs were further shown to express the *Brachyury* (*T*) gene as well as the *Scl* (Stem Cell Leukaemia) transcription factor [12, 13]. Later, studies on mouse embryos demonstrated the existence of the hemangioblast *in vivo* and indicated that it is found prominently in the posterior primitive streak [14]. It probably migrates from there to the yolk sac where the generation of blood, endothelial and vascular smooth muscle cells take place [14]. The existence of haemangioblast was also more recently documented in human with human ES cells [15] and *in vivo* in zebrafish [16].

Another concept of development proposes that a mature endothelial cell with haematopoietic potential, a haemogenic endothelium, give rise to blood cells. Several *in vitro* studies demonstrated that endothelial cells have the potential to generate blood cell lineages [17, 18]. In these studies, the authors isolated cells expressing both FLK1 and the endothelial marker VE-Cadherin and observed that these cells were able of *de novo* production of blood cells, marked by the expression of CD45. The generation of blood cells from endothelial progenitors was also demonstrated *in vivo* by Jaffredo and collaborators [19]. These authors specifically labelled endothelial cells in the avian embryo and observed that haematopoietic cells are later generated from these fluorescent endothelial progenitors.

More recently, a study by Lancrin and colleagues merged the haemangioblast and haemogenic endothelium theories into one linear model of development, in which the haemogenic endothelium is an intermediate stage during the generation of blood progenitors from the haemangioblast [20]. The presence of a haemogenic endothelium cell population was established both *in vitro* during ES cell differentiation as well as *in vivo*, in E7.5 mouse embryos [20-22]. In 2010, the generation of blood cells by haemogenic endothelium was directly visualised in embryos. This endothelial to haematopoietic transition (EHT) was observed both during murine [23] and zebrafish embryogenesis [24-26]. A schematic representation of the successive stages of haematopoietic commitment is presented in Figure 1.

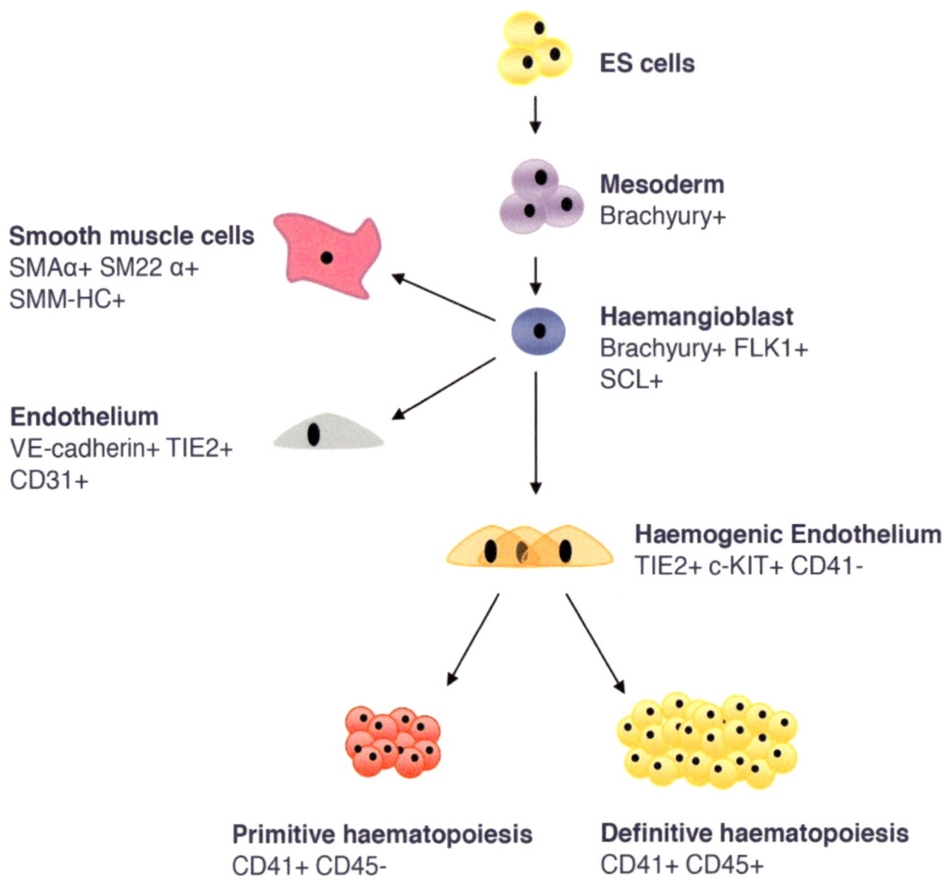

Figure 1. The process of generation of blood cells from the haemangioblast through a transient haemogenic endothelium cell population. Specific markers for each cell population are depicted. This figure has been adapted from [27].

2.4. Primitive erythrocytes – First blood cells in the embryo

Primitive erythrocytes, the first blood cells that emerge during embryogenesis in mammals, are large and nucleated. Their main function is to transport large quantities of oxygen to support the rapid growth of the embryo. It has been shown that these cells firstly appear *in vivo* in the yolk sac's blood islands around day E7.5 [28]. Although it was initially thought that these cells were nucleated, more recent studies have indicated that their nuclei are lost during the final days of mammalian gestation [29]. Studying primitive erythropoiesis remains challenging for several reasons; not only is the murine embryo around day 7 of gestation extremely small and difficult to access, but also there are no specific cell surface markers to specifically label this cell population. Therefore, transgenic mouse models were developed to directly address this later limitation.

Two mouse models used the expression of the same haemoglobin – ε-globin, as a reporter to track primitive erythropoiesis during ontogeny. In the first model [30], the authors coupled the ε-globin promoter to the KGFP (jellyfish-derived) fluorescent protein. With this new tool, they were able to isolate circulating primitive erythrocytes at different stages of murine embryogenesis and also to define the cell surface markers expressed by these cells between day E9.5 and E12.5 such as TER119, CD71, CD24, CD55 or CD147 [30]. In a second model, the fluorescent reporter used was a H2B-EGFP fusion protein [31]. This model allowed study of the complex process of maturation of primitive erythrocytes within the foetal liver [31]. More recently, the same group monitored the emergence of primitive erythrocytes at the very early stages of development *in vivo*, starting from day E6.75 and defined key pathways governing the emergence of this cellular lineage [32]. Other studies have more directly examined the cell signalling pathways supporting the emergence of primitive erythrocytes using both *in vivo* mouse models, as well as *in vitro* using the ES cell differentiation approach. In 2008, the importance of Wnt signalling for the emergence of primitive erythrocytes from FLK1 positive mesoderm was demonstrated [33]. Later, Cheng and co-workers expanded those findings and showed that not only the activation of Wnt pathway is crucial, but also the inhibition of Notch signalling is important for the emergence of primitive erythrocytes from FLK1 positive cells [34].

2.5. *In vivo* studies of the first haematopoietic stem cells (HSCs) – the foundation of the adult blood system

There are several definitive haematopoietic lineages generated during embryogenesis such as myeloid, lymphoid, definitive erythroid and haematopoietic stem cells. HSCs are the foundation of the blood system in the adult organism as these cells can differentiate towards all definitive haematopoietic cells.

HSCs arise first in the AGM, they possess the ability to self renew and, upon transplantation, they provide multilineage haematopoietic reconstitution [35]. Various studies, spanning several decades, aimed to characterise HSCs. In 1993, Huang and Auerbach reported that at E9.0 the murine yolk sac contains HSCs [36]. These cells, however, were unable to provide long-term multilineage haematopoietic reconstitution. A few months later, Muller and co-workers demonstrated that at day E10.5 the AGM region of the mouse contains fully functional HSCs – able to provide long-term haematopoietic reconstitution [7]. These findings were then expanded and HSCs were shown to emerge and expand within the anterior part of the AGM region [8]. It was also observed that definitive HSCs are present in the placenta, both in mouse [9, 37, 38] and human [10]. Interestingly, more recently mouse embryonic head tissues were shown to contain HSCs [39].

Several research groups investigated the cellular origin of haematopoietic stem cells *in vivo*. Zovein and co-workers (2008) demonstrated that HSCs emerge from the endothelium by performing lineage tracing experiment to specifically label either the endothelium or mesenchyme [21]. Furthermore, the emergence of putative haematopoietic cells from haemogenic endothelium has been visualised in the mouse embryonic aorta [23]. Similar results were obtained in zebrafish [24-26]. However whether these blood cells display any long-term repopulation activity remains to be directly assessed.

2.6. Molecular regulation of early embryonic haematopoiesis

Specific transcription factors regulate the developmental potential of different cells and progenitors during blood formation. In this section, we will discuss the role and function of a restricted set of these players.

One of the first genes implicated in mesoderm leading to blood development is the *Brachyury* gene. This transcription factor belongs to the T-box gene family [40] and is expressed by all nascent mesodermal cells [41]. Its expression is detected in murine embryo as early as E6.5 [42] and its deletion results in serious developmental defects and lethality by midgestation [42, 43]. In 2003, Fehling and co-workers generated a transgenic ES cell line in which GFP was targeted to the *Brachyury* locus. Further differentiation of this ES cell line allowed them to isolate mesodermal cells. The authors demonstrated that when combined with FLK1 (*Fetal Liver Kinase 1*) it was possible to separate three distinct populations corresponding to pre-mesoderm (negative for both FLK1 and *Brachyury*), mesodermal (positive for *Brachyury* only) and finally haemangioblastic, positive for both FLK1 and *Brachyury*, populations [13].

Flk1 encodes the receptor 2 for vascular endothelial growth factor (VEGF-R2) [44]. This gene is expressed by intra-and extra-embryonic mesoderm and later by endothelial cells in the vasculature [45, 46]. In 1995, Shalaby and colleagues demonstrated that in *Flk1* deficient embryos blood vessels and blood islands are not formed and that hardly any haematopoietic progenitors are present in these embryos [47]. As a result, *Flk1* deletion is embryonic lethal and embryos die between day E8.5 and E9.5 [47]. A few years later, they also determined, by evaluating the contribution of *Flk1* deficient ES cells to chimaeric mice, that the expression of *Flk1* is crucial for the migration of mesodermal progenitors from the intra-embryonic locations to the yolk sac [44]. This finding was later confirmed using *in vitro* differentiation of ES cells. In 1998 Choi and co-workers demonstrated that during ES cell differentiation, *Flk1* expression marks the blast colony-forming cell (BL-CFC) [11]. Later, it was shown that although Flk1 deficient ES cells are able to give rise to endothelial and haematopoietic lineages upon *in vitro* differentiation, they generate reduced number of blast colonies [48]. To date FLK1 remains, with *Brachyury*, the best marker of haemangioblast.

Etv2, a transcription factor of the *Ets* family, is another important regulator of haematopoietic specification. Murine embryos deficient for the expression of *Etv2* were shown to die at around E10.5 and to lack blood cells and vessels [49]. More recently, it was shown that *Etv2* is not required for the specification of primitive mesoderm, but is indispensable in the commitment of FLK1-positive mesoderm towards haematopoietic and endothelial programmes [50]. Using a transgenic *Etv2* ES cell line and mouse line it was shown that the expression of *Etv2* marks the endothelium and in particular haemogenic endothelial cell population [51]. In the absence of this transcription factor, both *in vivo* and *in vitro*, no haemogenic endothelium was observed [51]. Furthermore, it was demonstrated using a Cre-mediated deletion of *Etv2*, that this transcription factor is acting at the *Flk1* stage and that re-introduction of the *Scl* transcription factor in Etv2-deficient ES cells can fully rescue the haematopoietic potential of these cells [52].

The transcription factor *Scl* (Stem Cell Leukaemia Factor) was originally identified due to its involvement in chromosomal translocation in T-cell leukaemias [53] and was later demonstrated by both *in vitro* and *in vivo* studies to play a significant role during embryonic blood development. Murine embryos lacking this transcription factor do not develop neither primitive nor definitive haematopoietic cells and die by E9.5 [54, 55]. In addition, *Scl-/-*ES cells do not generate blast colonies (Robertson et al., 2000) or any haematopoietic cells [56]. Blast colony forming cells (BL-CFCs) were shown to express Scl during *in vitro* ES cell differentiation [12]. More recently it was demonstrated that *Scl* is critical for the generation of the haemogenic endothelium [20].

SOX7, with SOX17 and SOX18, form the F-subgroup of SRY-related (HMG-box) family of transcription factors [57]. During embryonic development SOX7 transcripts are detected in various tissues such as brain, heart, lung, kidney and spleen [58]. SOX7 and SOX18 knock-downs performed in zebrafish and *Xenopus* embryos revealed critical roles of these transcription factors in cardiogenesis and vasculogenesis [59-61]. Recently, Wat and colleagues developed a mouse model lacking the expression of Sox7 that is embryonic lethal at E10.5 due to cardiovascular abnormalities [62]. SOX7 was recently shown to be also implicated in early stages of blood development. SOX7 expression is upregulated at the haemangioblast stage and transiently expressed in the first CD41 – positive blood progenitors emerging from the FLK1-positive haemangioblasts [63]. Its enforced expression in haematopoietic progenitors, marked by the expression of CD41, results in the arrest of haematopoietic differentiation of these cells, a property shared by Sox18 but not Sox17 [64]. Recently, it was also reported that SOX7 is expressed at the haemogenic endothelium stage, where it regulates the expression of the endothelial marker VE-Cadherin [65].

The transcription factor RUNX1, encoded by the *Runx1/AML1* gene is considered a master regulator of definitive haematopoiesis. Indeed *Runx1* deficient embryos completely lack definitive haematopoietic cells [66]. The deletion of *Runx1* gene is embryonic lethal by E11.5 and E12.5 of gestation, and these embryos present multiple haemorrhages [67]. *Runx1* was also shown to be critical *in vitro*. *Runx1-/-*ES cells generate only a few blast colonies and these are restricted to primitive haematopoietic programme [68]. Furthermore the kinetic of the development of the haematopoietic system has been shown to be dependent of a gene dosage effect of *Runx1* [69]. Finally, *Runx1* is essential for the formation of haematopoietic progenitors from the haemogenic endothelium [20, 70, 71]. This critical role of RUNX1 in the endothelial to haematopoietic transition has spurred efforts to identify and characterize its direct transcriptional targets [72-74].

Although, numerous molecular regulators of haematopoietic specification have been identified, it is likely that many others remain to be discovered. In addition, the events they regulate and how they interact to orchestrate blood development remain largely unknown. The specific requirement for several of these regulators is depicted in Figure 2.

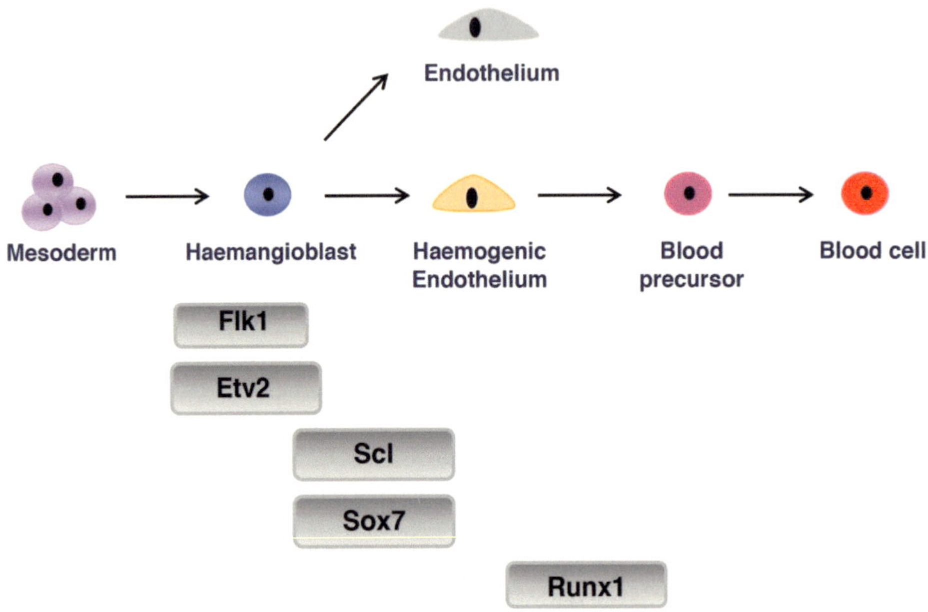

Figure 2. Molecular regulation of early embryonic haematopoiesis. The stages of blood development where the function of the different genes is critical are indicated.

3. Methods and procols to study haematopoitic development during es cell differentiation

ES cell differentiation provides a relatively easy and accessible system to study early embryonic haematopoiesis. Using well-defined protocols, it is possible to effectively study the events happening *in vivo* using this *in vitro* approach. This experimental system was shown to recapitulate the early *in vivo* events of development of the haematopoietic system.

3.1. Mouse embryonic fibroblasts (MEFs)

There are several methods to keep ES cells undifferentiated. One of them consists of growing ES cells on mouse embryonic fibroblasts (MEFs). Before working with ES cells, it is recommended to prepare a good stock of MEFs to be used as feeder cell layer. For that, wild type ICR or DR4 (resistant to four drugs) [75] MEFs are harvested from E14.5 embryos and cultured in Iscove's modified Dulbecco medium (IMDM, Lonza) supplemented with 50 μg/ml penicillin-streptomicin (Gibco), 2mM L-Glutamine, 10% of FCS (PAA Laboratories) and $1,5x10^{-4}$ monothioglycerol (MTG, Sigma) under low oxygen conditions. When amplified, MEFs are harvested (TrypLE, Invitrogen) and irradiated at 30Gy to stop the cells proliferation. The cells should be frozen at around 1 million cells per ml of IMDM supplemented with 50% FCS and 10% of dimethyl sulfoxide (DMSO). Cells should be stored at-80 ^{0}C. Thawed MEFs should be replated in one six-well plate previously coated with gelatine and let to adhere to the plastic wells overnight. Upon microscopic examination, MEFs should cover the entire surface of the cell-culture dish and be ready to be seeded with ES cells.

3.2. ES cell culture

ES cells are cultured on irradiated MEFs in a media constituted of Dulbeco's modified Eagle Medium (DMEM, Gibco) supplemented with 50 µg/ml penicillin-streptomycin, 2mM L-Glutamine, 15% FCS (PAA Laboratories), 2% Leukaemia Inhibitory Factor (LIF) (conditioned medium from LIF-generating cell line, see [76]) or 50 units of recombinant ESGRO LIF/ml (Millipore) and $1,5x10^4$ MTG (Sigma). Leukaemia inhibitory factor (LIF) – is a cytokine inhibiting differentiation. ES cells, when cultured on MEFs feeder cell layer in the presence of LIF remain undifferentiated. Upon microscopic observation they form tightly associated clusters of cells that are bright and shiny in appearance (Fig. 3A).

3.3. Generation of embryoid bodies

Embryoid bodies (EBs) are three-dimensional structures spontaneously generated by ES cells during differentiation. They contain precursors for the three primary germ layers ectoderm, endoderm and mesoderm. Two passages on gelatine are performed to remove the MEFs that would hamper ES cells differentiation. The first passage is performed in DMEM-ES media (described above), whereas for the second passage DMEM is replaced with IMDM. The ES cells are then harvested by trypsinisation and seeded into liquid cultures in non-tissue culture Petri dishes (Sterilin) in differentiation medium containing: IMDM supplemented with 15% FCS serum selected for differentiation (PAA Laboratories), 2mM L-Glutamine, 180 µg/ml transferring (Roche), 25 µg/ml Ascorbic Acid (AA, Sigma) and $4,6x10^{-4}$ MTG. The density of cell seeding should be adjusted in function of the day at which the cultures will be harvested, varying from $1,5x10^4$ cells/ml (for day 4-6) up to $3,0 \times 10^4$ cells/ml (for days 2.5-3.5). 10-20ml of "Differentiation medium" should be used per one Petri dish.

By performing two passages on gelatine and removing feeder cell layer and LIF, ES cells become primed for differentiation and formation of three-dimensional embryoid bodies in liquid culture (Figure 3B). Early EBs contain precursors for the three primary germ layers. By day 7, hemoglobinisation can be observed as red areas present within the EBs (Figure 3C). This system is versatile and allows to access and study in details several subsequent stages of blood development such as the emergence of haemangioblast, production of blast colonies and the development of primitive and definitive blood precursors (Figure 4).

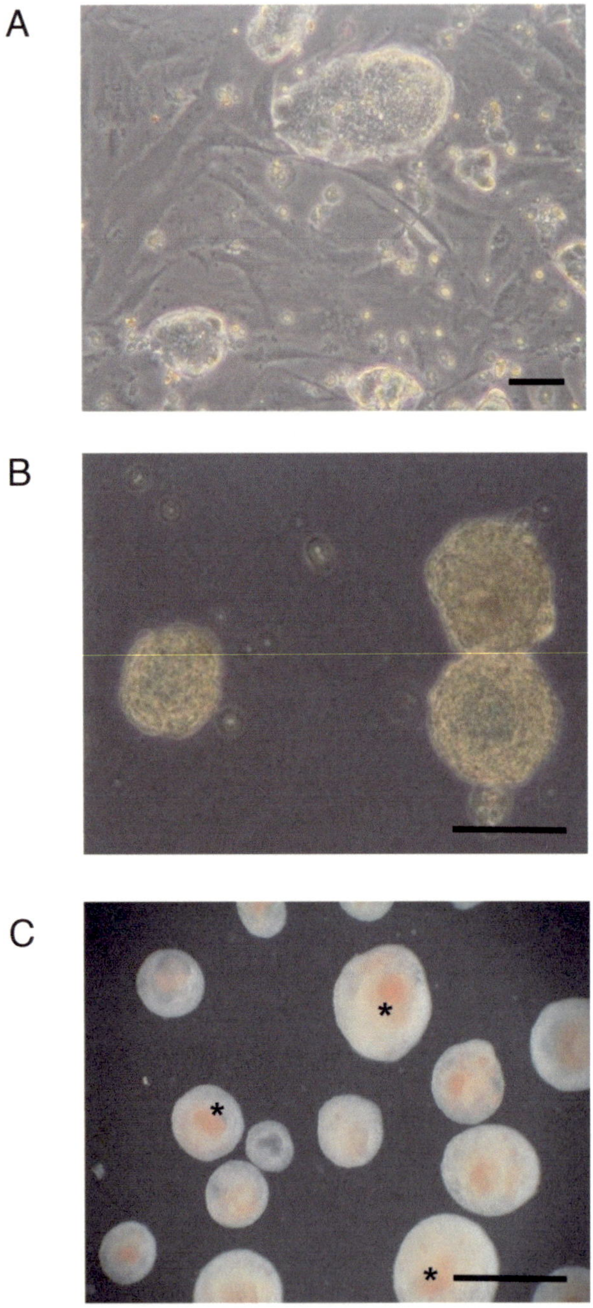

Figure 3. Morphology of ES cells and embryoid bodies (EBs). A) Typical appearance of ES cell in culture. Cells form bright, shiny adherent colonies and are cultured on a layer of MEFs feeder cell layer. B) Typical appearance of EBs in culture. These three-dimensional structures are formed during ES cell differentiation and contain precursors for three primary germ cell layers. C) Embryoid bodies at day 7 of differentiation containing haemoglobin (indicated with asterisks). Scale bar 300μm.

http://dx.doi.org/10.5772/57607

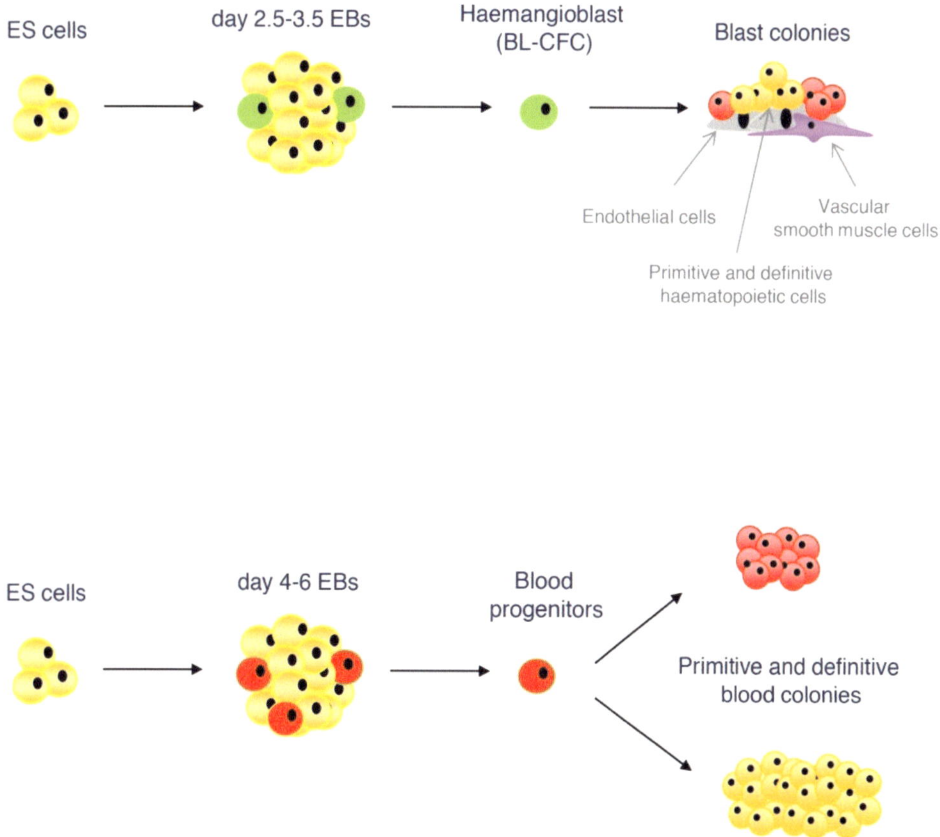

Figure 4. The ES/EB differentiation system. Upon differentiation of ES cells, three-dimensional structures, called embryoid bodies (EBs) are formed. Sorting EBs at day 2.5-3.5 for the expression of FLK1 enrich for BL-CFC, the *in vitro* equivalent of the haemangioblast. Upon culture, the BL-CFCs generate blast colonies that contain precursors for haematopoietic, endothelial and vascular smooth muscle cells. Day 4-6 EBs contain haematopoietic progenitors that can give rise to various primitive and definitive haematopoietic colonies.

3.4. Blast colonies

There are two alternative approaches to study the development of blast colonies; liquid culture on gelatine or semi-solid culture in methylcellulose. Liquid culture on gelatine facilitates harvesting of the cells for flow cytometry analysis or time-lapse imaging techniques. Alternatively, the semi-solid culture is a clonogenic assay allowing the growth of of individual blast colonies and their quantification. Individual blast colonies can then further be isolated and their haematopoietic potential assessed. To perform blast colony culture, EBs are harvested between day 2.5 and 3.5 and the cells are enriched for haemangioblast by sorting for the expression of FLK1, the receptor 2 for VEGF. Isolated FLK1 positive cells are then replated on gelatinised cell culture plates at the density of 8,5 x 10^4 cell/$10cm^2$ in a medium containing IMDM supplemented with 10% FCS, 2mM L-Glutamine mM, 180 µg/ml transferrin, 25 µg/ml ascorbic acid, 4.6 x 10-4 M MTG, 15% endothelial cell line-D4T conditioned medium [11], 10 ng/ml interleukin 6 (IL-6) (Peprotech) and 5 ng/ml vascular endothelial growth factor (VEGF, Peprotech). For semi-solid cultures, FLK1 positive cells are seeded in 35 mm x 10 mm (BD Falcon) dishes at a density of 1.5×10^4 cells/ml in IMDM medium supplemented with 10% FCS, 2mM L-Glutamine mM, 180 μg/ml transferrin, 25 μg/ml ascorbic acid, 4.6 x 10-4 M MTG, 15% endothelial cell line-D4T conditioned medium [11], 10 ng/ml interleukin 6 (IL-6) (Peprotech) and 5 ng/ml vascular endothelial growth factor (VEGF, Peprotech). The medium is additionally supplemented with 10 g/L methylcellulose (dissolved in IMDM, Alfa-Aesar). Blast colonies are scored 3-4 days after replating. Plating 3.0×10^4 cells per 1 ml of semi-solid medium should result in the formation of around 300-400 of blast colonies.

Sorting EBs between day 2.5 and 3.5 for the expression of FLK1 enrich for blast colony forming cell (BL-CFCs) [11, 13]. During the formation of blast colonies, several distinct morphological stages can be observed. First, there is the formation of tight clusters of adherent cells and later, single round cells emerging from these tight clusters. These cells then proliferate (Figure 5A). Alongside these morphological changes, there is first an upregulation of endothelial markers, such as TIE2 and VE-Cadherin during the formation of the tight clusters. Later the expression of these endothelial markers is gradually downregulated whereas the expression of haematopoietic markers (such as CD41 and then CD45) is upregulated (Figure 5B). This correlates with the emergence of round floating–haematopoietic-cells.

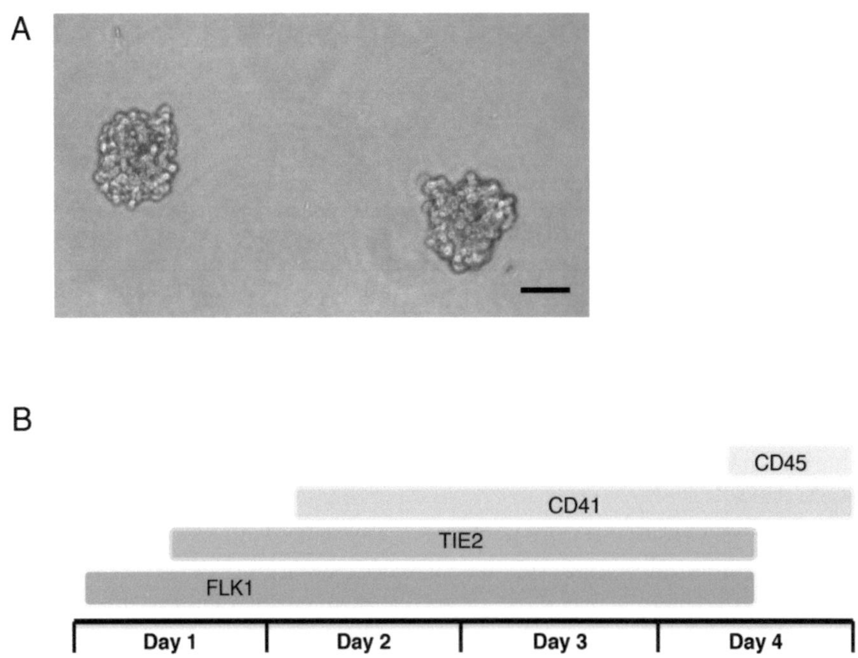

Figure 5. The development of the blast colony during four days of differentiation. A) Microscopic image showing representative blast colonies. Scale bar 100μm. B) Schematic representation of the different cell surface markers expressed on cells during blast colony differentiation.

It was recently shown that the haemangioblast progenitor can also generate vascular smooth muscle cells in addition to endothelial and haematopoietic cells [14, 77, 78]. Accordingly, Yamashita and co-workers also established that FLK1 positive cells can differentiate towards endothelial and mural cells [78]. Smooth muscle cells appear, under the microscope as large flat adherent cells. These cells express smooth muscle specific genes such as smooth muscle actin α, transgelin, calponin, and smooth muscle myosin heavy chain [79]. A typical immunostaining of α-smooth muscle actin is presented in figure 6A. We have recently analysed in more detail the generation of smooth muscle cells and showed that these cells are largely generated independently from the haemogenic endothelium [80].

A

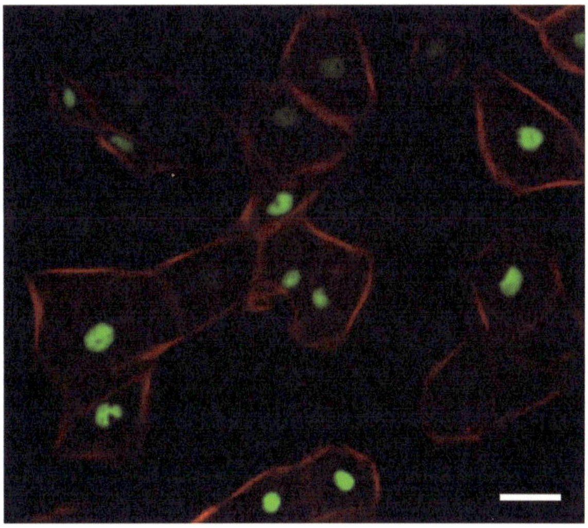

Figure 6. The differentiation of ES cells towards smooth muscle cells. Representative image of smooth muscle cells stained with SMAα-Cy3 antibody. These cells also carry a transgenic BAC containing H2B-VENUS cDNA under the control of the α-smooth muscle actin transcriptional regulatory elements. Scale bar 100μm.

3.5. Culture of haemogenic endothelium

During blast colony development *in vitro*, tight clusters of adherent cells that represent haemogenic endothelium are observed after around two days. This population of cells corresponds to an intermediate stage in the formation of blood cells from the haemangioblast. It was recently shown that these cells express TIE2 and VE-cadherin (markers of endothelial cells), c-KIT (expressed by both haematopoietic and endothelial cells) and are negative for CD41 (first haematopoietic marker) expression [20]. To isolate haemogenic endothelium, cells in day 2 liquid blast cultures are harvested and FACS sorted for the expression of these markers [20]. It is also possible to isolate a more advanced haemogenic endothelium cell population, positive for the expression of CD41 [70]. The cells are next plated onto gelatin-coated plates at a density of 1.2 x 10^5 cells/cm^2 in IMDM medium supplemented with 10% FCS, 2 mM L-Glutamine, 180 μg/ml transferrin, 25 μg/ml ascorbic acid, 4.6 x 10-4 M MTG, 10 ng/ml Oncostatin M (R&D Systems) and 1 ng/ml bFGF (Peprotech). Cells are grown in standard culture conditions. Upon culture, at least 1-2% of these cells generate primitive and definitive haematopoietic cells [20]. During this process, the haemogenic endothelial cell population (TIE2+,c-KIT+,CD41-) cells gradually acquire the expression of CD41. The cells then progress further and lose their expression of endothelial markers.

3.6. Haematopoietic colonies assays

To evaluate the presence of haematopoietic colonies, EBs should be harvested at day 4,5 or 6 and trypsinised. Cells from the EBs can be then directly used or alternatively sorted for the expression of a marker of haematopoietic progenitors, such as for example CD41. Approximately 3.0×10^4 unsorted cells should be plated in 35 mm x 10 mm (BD Falcon) dishes in 1 ml of semisolid medium containing IMDM, 15% plasma derived serum (PDS) (Antech), 10% protein free hybridoma medium (PFHM, Gibco), 2mM L-Glutamine, 180 μg/ml transferrin, 25 μg/ml ascorbic acid, 4.6 x 10-4 M MTG and cytokines such as: 1% c-KIT ligand supernatant, 1% interleukin 3 supernatant (IL-3) (see [76]), 1μg/ml GM-CSF, 1% thrombopoietin conditioned media, 10 ng/ml IL-6 (Peprotech), 10 ng/ml macrophage colony stimulating factor (M-CSF), 5 ng/ml IL-11 (R&D Systems) and 4 U/ml of Erythropoietin (Ortho-Biotech) and 10 g/L methylcellulose (dissolved in IMDM, Alfa-Aesar). Haematopoietic colonies are assessed and scored based on their morphology. Primitive erythroid colonies are scored at day 5, whereas definitive haematopoietic colonies are usually enumerated 8 days after replating. Morphologic landmarks are used to distinguish the different types of haematopoietic colonies. Haematopoietic progenitors can also be cultured in liquid conditions to allow easier access of the cells for subsequent flow cytometry analysis or cytospin assays. For that cells should be seeded at a density of 2.0×10^6 /ml in ultra low-adherence tissue culture plates (Costar) in the haematopoietic medium described above with methylcellulose being replaced with IDMD medium.

The onset of emergence of primitive erythroid cells is observed within EBs by day 4 of differentiation [81]. Definitive erythroid and macrophage precursors appear shortly after and are followed by mast cells and multilineage precursors [81]. Primitive colonies appear around day 4 of culture. These colonies are round, compact and bright red in colour. By day 6-7 of culture, morphologically distinguishable definitive haematopoietic colonies are detected (Figure 7).

4. Conclusions and future directions

In this chapter, we have reviewed recent progress in our understanding of the development of the haematopoietic system. We have emphasized the critical role that the use of ES cells, in particular murine ES cells, has played in these recent advances. ES cells have been instrumental to identify and characterise the elusive haemangioblast. More recently, this model system allowed the merging of the two conflicting theories of the origin of blood cells (haemangioblast and haemogenic endothelium) in a single linear model of development. In addition the precise roles and requirements of many critical regulators of this process have been to a large extent elucidated using this approach.

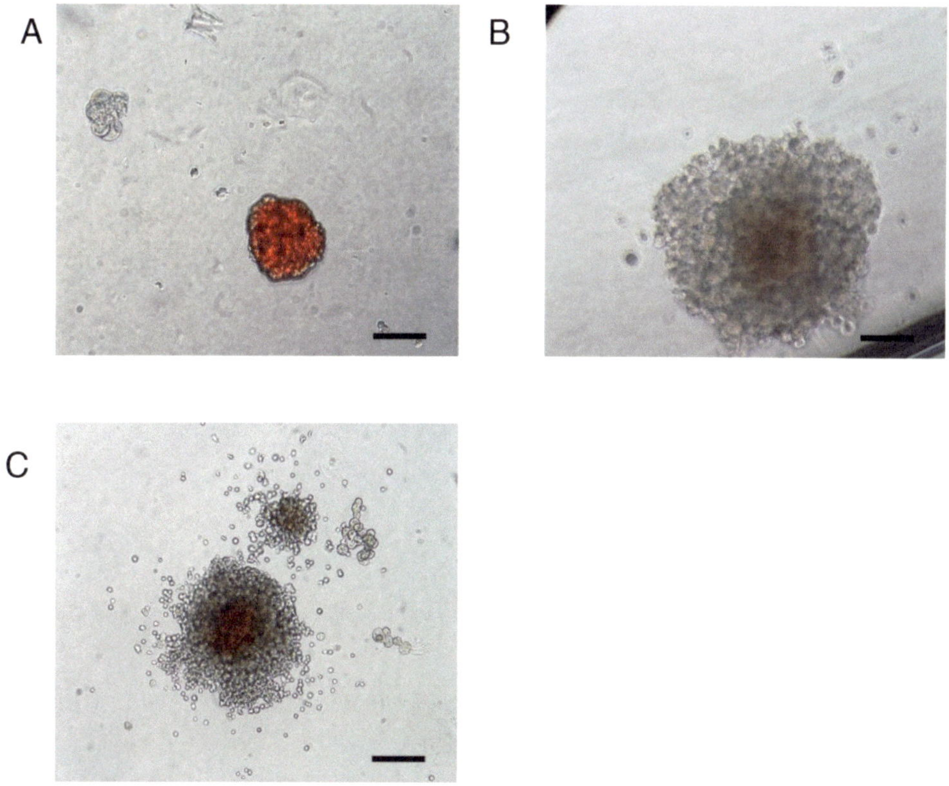

Figure 7. Examples of haematopoietic colonies obtained during ES cell differentiation A) Primitive erythroid colony – red in colour, compact and relatively small. B) Definitive myeloid colony – bigger in size, white and looser cells. C) Mixed haematopoietic colony. Scale bar 50μm.

Although only very succinctly discussed in this chapter, this system is also very amenable to examination of the cell signalling pathways that support the development of normal haematopoiesis [33, 34, 82-84]. Interestingly these pathways are also implicated in leukaemogenesis [85, 86]. Finally with the advent of novel human ES cells or human iPS cells [87], that recapitulate better the ground state and are easier to work with, and development of new methods facilitating genome editing [88-92], this experimental system is very likely to be instrumental for delivering new advances in our understanding of human haematopoietic development, that is otherwise very difficult to study *in vivo*.

Acknowledgements

We apologize to the many colleagues whose work could not be cited owing to space constraints. Our work is supported by Cancer Research UK (CRUK), Leukaemia and Lymphoma research (LLR) and the Biotechnology and Biological Sciences Research Council (BBSRC).

Author details

Monika Stefanska[1,2], Valerie Kouskoff[3] and Georges Lacaud[1]

1 Cancer Research UK Manchester Institute, Stem Cell Biology Group, University of Manchester, Manchester, United Kingdom

2 Jagiellonian University, Faculty of Biochemistry, Biophysics and Biotechnology, Kraków, Poland

3 Cancer Research UK Manchester Institute, Stem Cell Haematopoiesis Group, University of Manchester, Manchester, United Kingdom

References

[1] Keller, G.M., *In vitro differentiation of embryonic stem cells.* Curr Opin Cell Biol, 1995. 7(6): p. 862-9.

[2] Smith, A.G., *Embryo-derived stem cells: of mice and men.* Annu Rev Cell Dev Biol, 2001. 17: p. 435-462.

[3] Sabin, F.R., *Studies on the origin of blood-vessels and of red blood-corpuscles as seen in the living blastoderm of chicks during the second day of incubation.* Contributions to Embryology 1920. 9(9): p. 215-262.

[4] Maximow, A.A., *Relation of blood cells to connective tissues and endothelium.* Physiological Reviews, 1924. 4(4): p. 533-563.

[5] Haar, J.L. and G.A. Ackerman, *A phase and electron microscopic study of vasculogenesis and erythropoiesis in the yolk sac of the mouse.* Anat Rec, 1971. 170(2): p. 199-223.

[6] Palis, J., et al., *Development of erythroid and myeloid progenitors in the yolk sac and embryo proper of the mouse.* Development, 1999. 126(22): p. 5073-84.

[7] Muller, A.M., et al., *Development of hematopoietic stem cell activity in the mouse embryo.* Immunity, 1994. 1(4): p. 291-301.

[8] Medvinsky, A. and E. Dzierzak, *Definitive hematopoiesis is autonomously initiated by the AGM region.* Cell, 1996. 86(6): p. 897-906.

[9] Gekas, C., et al., *The placenta is a niche for hematopoietic stem cells.* Developmental Cell, 2005. 8(3): p. 365-75.

[10] Robin, C., et al., *Human placenta is a potent hematopoietic niche containing hematopoietic stem and progenitor cells throughout development.* Cell Stem Cell, 2009. 5(4): p. 385-95.

[11] Choi, K., et al., *A common precursor for hematopoietic and endothelial cells.* Development, 1998. 125(4): p. 725-32.

[12] Faloon, P., et al., *Basic fibroblast growth factor positively regulates hematopoietic development.* Development, 2000. 127(9): p. 1931-41.

[13] Fehling, H.J., et al., *Tracking mesoderm induction and its specification to the hemangioblast during embryonic stem cell differentiation.* Development, 2003. 130(17): p. 4217-27.

[14] Huber, T.L., et al., *Haemangioblast commitment is initiated in the primitive streak of the mouse embryo.* Nature, 2004. 432(7017): p. 625-30.

[15] Kennedy, M., et al., *Development of the hemangioblast defines the onset of hematopoiesis in human ES cell differentiation cultures.* Blood, 2007. 109(7): p. 2679-87.

[16] Vogeli, K.M., et al., *A common progenitor for haematopoietic and endothelial lineages in the zebrafish gastrula.* Nature, 2006. 443(7109): p. 337-9.

[17] Nishikawa, S.I., et al., *Progressive lineage analysis by cell sorting and culture identifies FLK1+VE-cadherin+cells at a diverging point of endothelial and hemopoietic lineages.* Development, 1998. 125(9): p. 1747-57.

[18] Nishikawa, S.I., et al., *In vitro generation of lymphohematopoietic cells from endothelial cells purified from murine embryos.* Immunity, 1998. 8(6): p. 761-9.

[19] Jaffredo, T., et al., *Intraaortic hemopoietic cells are derived from endothelial cells during ontogeny.* Development, 1998. 125(22): p. 4575-83.

[20] Lancrin, C., et al., *The haemangioblast generates haematopoietic cells through a haemogenic endothelium stage.* Nature, 2009. 457(7231): p. 892-5.

[21] Zovein, A.C., et al., *Fate tracing reveals the endothelial origin of hematopoietic stem cells.* Cell Stem Cell, 2008. 3(6): p. 625-36.

[22] Eilken, H.M., S. Nishikawa, and T. Schroeder, *Continuous single-cell imaging of blood generation from haemogenic endothelium.* Nature, 2009. 457(7231): p. 896-900.

[23] Boisset, J.C., et al., *In vivo imaging of haematopoietic cells emerging from the mouse aortic endothelium.* Nature, 2010. 464(7285): p. 116-20.

[24] Kissa, K. and P. Herbomel, *Blood stem cells emerge from aortic endothelium by a novel type of cell transition.* Nature, 2010. 464(7285): p. 112-5.

[25] Bertrand, J.Y., et al., *Haematopoietic stem cells derive directly from aortic endothelium during development.* Nature, 2010. 464(7285): p. 108-11.

[26] Lam, E.Y., et al., *Live imaging of Runx1 expression in the dorsal aorta tracks the emergence of blood progenitors from endothelial cells.* Blood, 2010. 116(6): p. 909-14.

[27] Lancrin, C., et al., *Blood cell generation from the hemangioblast.* J Mol Med (Berl), 2010. 88(2): p. 167-72.

[28] Palis, J., K.E. McGrath, and P.D. Kingsley, *Initiation of hematopoiesis and vasculogenesis in murine yolk sac explants.* Blood, 1995. 86(1): p. 156-63.

[29] Kingsley, P.D., et al., *Yolk sac-derived primitive erythroblasts enucleate during mammalian embryogenesis.* Blood, 2004. 104(1): p. 19-25.

[30] Fraser, S.T., J. Isern, and M.H. Baron, *Maturation and enucleation of primitive erythroblasts during mouse embryogenesis is accompanied by changes in cell-surface antigen expression.* Blood, 2007. 109(1): p. 343-52.

[31] Isern, J., et al., *The fetal liver is a niche for maturation of primitive erythroid cells.* Proc Natl Acad Sci U S A, 2008. 105(18): p. 6662-7.

[32] Isern, J., et al., *Single-lineage transcriptome analysis reveals key regulatory pathways in primitive erythroid progenitors in the mouse embryo.* Blood, 2011. 117(18): p. 4924-34.

[33] Nostro, M.C., et al., *Wnt, activin, and BMP signaling regulate distinct stages in the developmental pathway from embryonic stem cells to blood.* Cell Stem Cell, 2008. 2(1): p. 60-71.

[34] Cheng, X., et al., *Numb mediates the interaction between Wnt and Notch to modulate primitive erythropoietic specification from the hemangioblast.* Development, 2008. 135(20): p. 3447-58.

[35] Lemischka, I.R., *Clonal, in vivo behavior of the totipotent hematopoietic stem cell.* Semin Immunol, 1991. 3(6): p. 349-55.

[36] Huang, H. and R. Auerbach, *Identification and characterization of hematopoietic stem cells from the yolk sac of the early mouse embryo.* Proc Natl Acad Sci U S A, 1993. 90(21): p. 10110-4.

[37] Ottersbach, K. and E. Dzierzak, *Analysis of the mouse placenta as a hematopoietic stem cell niche.* Methods Mol Biol., 2009. 538: p. 335-46.

[38] Rhodes, K.E., et al., *The emergence of hematopoietic stem cells is initiated in the placental vasculature in the absence of circulation.* Cell Stem Cell, 2008. 2(3): p. 252-63.

[39] Li, Z., et al., *Mouse embryonic head as a site for hematopoietic stem cell development.* Cell Stem Cell, 2012. 11(5): p. 663-75.

[40] Kispert, A., B. Koschorz, and B.G. Herrmann, *The T protein encoded by Brachyury is a tissue-specific transcription factor.* EMBO J, 1995. 14(19): p. 4763-72.

[41] Wilkinson, D.G., S. Bhatt, and B.G. Herrmann, *Expression pattern of the mouse T gene and its role in mesoderm formation.* Nature, 1990. 343(6259): p. 657-9.

[42] Kispert, A. and B.G. Herrmann, *Immunohistochemical analysis of the Brachyury protein in wild-type and mutant mouse embryos.* Dev Biol, 1994. 161(1): p. 179-93.

[43] Yanagisawa, K.O., H. Fujimoto, and H. Urushihara, *Effects of the brachyury (T) mutation on morphogenetic movement in the mouse embryo.* Dev Biol, 1981. 87(2): p. 242-8.

[44] Shalaby, F., et al., *A requirement for Flk1 in primitive and definitive hematopoiesis and vasculogenesis.* Cell, 1997. 89(6): p. 981-90.

[45] Yamaguchi, T.P., et al., *flk-1, an flt-related receptor tyrosine kinase is an early marker for endothelial cell precursors.* Development, 1993. 118(2): p. 489-98.

[46] Millauer, B., et al., *High affinity VEGF binding and developmental expression suggest Flk-1 as a major regulator of vasculogenesis and angiogenesis.* Cell, 1993. 72(6): p. 835-46.

[47] Shalaby, F., et al., *Failure of blood-island formation and vasculogenesis in Flk-1-deficient mice.* Nature, 1995. 376(6535): p. 62-6.

[48] Schuh, A.C., et al., *In vitro hematopoietic and endothelial potential of flk-1(-/-) embryonic stem cells and embryos.* Proc Natl Acad Sci U S A, 1999. 96(5): p. 2159-64.

[49] Lee, D., et al., *ER71 acts downstream of BMP, Notch, and Wnt signaling in blood and vessel progenitor specification.* Cell Stem Cell, 2008. 2(5): p. 497-507.

[50] Kataoka, H., et al., *Etv2/ER71 induces vascular mesoderm from Flk1+PDGFRalpha+primitive mesoderm.* Blood, 2011. 118(26): p. 6975-86.

[51] Wareing, S., et al., *ETV2 expression marks blood and endothelium precursors, including hemogenic endothelium, at the onset of blood development.* Dev Dyn, 2012. 241(9): p. 1454-64.

[52] Wareing, S., et al., *The Flk1-Cre-mediated deletion of ETV2 defines its narrow temporal requirement during embryonic hematopoietic development.* Stem Cells, 2012. 30(7): p. 1521-31.

[53] Begley, C.G., et al., *The gene SCL is expressed during early hematopoiesis and encodes a differentiation-related DNA-binding motif.* Proc Natl Acad Sci U S A, 1989. 86(24): p. 10128-32.

[54] Robb, L., et al., *Absence of yolk sac hematopoiesis from mice with a targeted disruption of the scl gene.* Proc Natl Acad Sci U S A, 1995. 92(15): p. 7075-9.

[55] Shivdasani, R.A., E.L. Mayer, and S.H. Orkin, *Absence of blood formation in mice lacking the T-cell leukaemia oncoprotein tal-1/SCL.* Nature, 1995. 373(6513): p. 432-4.

[56] Porcher, C., et al., *The T cell leukemia oncoprotein SCL/tal-1 is essential for development of all hematopoietic lineages.* Cell, 1996. 86(1): p. 47-57.

[57] Bowles, J., G. Schepers, and P. Koopman, *Phylogeny of the SOX family of developmental transcription factors based on sequence and structural indicators.* Dev Biol, 2000. 227(2): p. 239-55.

[58] Takash, W., et al., *SOX7 transcription factor: sequence, chromosomal localisation, expression, transactivation and interference with Wnt signalling.* Nucleic Acids Res, 2001. 29(21): p. 4274-83.

[59] Zhang, C., T. Basta, and M.W. Klymkowsky, *SOX7 and SOX18 are essential for cardiogenesis in Xenopus.* Dev Dyn, 2005. 234(4): p. 878-91.

[60] Cermenati, S., et al., *Sox18 and Sox7 play redundant roles in vascular development.* Blood, 2008. 111(5): p. 2657-66.

[61] Pendeville, H., et al., *Zebrafish Sox7 and Sox18 function together to control arterial-venous identity.* Dev Biol, 2008. 317(2): p. 405-16.

[62] Wat, M.J., et al., *Mouse model reveals the role of SOX7 in the development of congenital diaphragmatic hernia associated with recurrent deletions of 8p23.1.* Hum Mol Genet, 2012. 21(18): p. 4115-25.

[63] Gandillet, A., et al., *Sox7-sustained expression alters the balance between proliferation and differentiation of hematopoietic progenitors at the onset of blood specification.* Blood, 2009. 114(23): p. 4813-22.

[64] Serrano, A.G., et al., *Contrasting effects of Sox17-and Sox18-sustained expression at the onset of blood specification.* Blood, 2010. 115(19): p. 3895-8.

[65] Costa, G., et al., *SOX7 regulates the expression of VE-cadherin in the haemogenic endothelium at the onset of haematopoietic development.* Development, 2012. 139(9): p. 1587-98.

[66] Okuda, T., et al., *AML1, the target of multiple chromosomal translocations in human leukemia, is essential for normal fetal liver hematopoiesis.* Cell, 1996. 84(2): p. 321-30.

[67] Wang, Q., et al., *Disruption of the Cbfa2 gene causes necrosis and hemorrhaging in the central nervous system and blocks definitive hematopoiesis.* Proc Natl Acad Sci U S A, 1996. 93(8): p. 3444-9.

[68] Lacaud, G., et al., *Runx1 is essential for hematopoietic commitment at the hemangioblast stage of development in vitro.* Blood, 2002. 100(2): p. 458-66.

[69] Lacaud, G., et al., *Haploinsufficiency of Runx1 results in the acceleration of mesodermal development and hemangioblast specification upon in vitro differentiation of ES cells.* Blood, 2004. 103(3): p. 886-9.

[70] Sroczynska, P., et al., *The differential activities of Runx1 promoters define milestones during embryonic hematopoiesis.* Blood, 2009. 114(26): p. 5279-89.

[71] Chen, M.J., et al., *Runx1 is required for the endothelial to haematopoietic cell transition but not thereafter.* Nature, 2009. 457(7231): p. 887-91.

[72] Ferreras, C., et al., *Identification and characterization of a novel transcriptional target of RUNX1/AML1 at the onset of hematopoietic development.* Blood, 2011. 118(3): p. 594-7.

[73] Lancrin, C., et al., *GFI1 and GFI1B control the loss of endothelial identity of hemogenic endothelium during hematopoietic commitment.* Blood, 2012. 120(2): p. 314-22.

[74] Tanaka, Y., et al., *The transcriptional programme controlled by Runx1 during early embryonic blood development.* Dev Biol, 2012. 366(2): p. 404-19.

[75] Tucker, K.L., et al., *A transgenic mouse strain expressing four drug-selectable marker genes.* Nucleic Acids Res, 1997. 25(18): p. 3745-6.

[76] Sroczynska, P., et al., *In vitro differentiation of mouse embryonic stem cells as a model of early hematopoietic development.* Methods Mol Biol, 2009. 538: p. 317-34.

[77] Ema, M., et al., *Combinatorial effects of Flk1 and Tal1 on vascular and hematopoietic development in the mouse.* Genes Dev, 2003. 17(3): p. 380-93.

[78] Yamashita, J., et al., *Flk1-positive cells derived from embryonic stem cells serve as vascular progenitors.* Nature, 2000. 408(6808): p. 92-6.

[79] Owens, G.K., *Regulation of differentiation of vascular smooth muscle cells.* Physiol Rev, 1995. 75(3): p. 487-517.

[80] Stefanska, M., et al., *Smooth muscle cells largely develop independently of functional hemogenic endothelium.* Stem Cell Res, 2013. 12(1): p222-232

[81] Keller, G., et al., *Hematopoietic commitment during embryonic stem cell differentiation in culture.* Mol Cell Biol, 1993. 13(1): p. 473-86.

[82] Pearson, S., et al., *The stepwise specification of embryonic stem cells to hematopoietic fate is driven by sequential exposure to Bmp4, activin A, bFGF and VEGF.* Development, 2008. 135(8): p. 1525-35.

[83] Holley, R.J., et al., *Influencing hematopoietic differentiation of mouse embryonic stem cells using soluble heparin and heparan sulfate saccharides.* J Biol Chem, 2011. 286(8): p. 6241-52.

[84] Baldwin, R.J., et al., *A developmentally regulated heparan sulfate epitope defines a subpopulation with increased blood potential during mesodermal differentiation.* Stem Cells, 2008. 26(12): p. 3108-18.

[85] Barker, N. and H. Clevers, *Mining the Wnt pathway for cancer therapeutics.* Nat Rev Drug Discov, 2006. 5(12): p. 997-1014.

[86] Clevers, H., *Wnt/beta-catenin signaling in development and disease.* Cell, 2006. 127(3): p. 469-80.

[87] Gafni, O., et al., *Derivation of novel human ground state naive pluripotent stem cells.* Nature, 2013.

[88] Wang, H., et al., *One-step generation of mice carrying mutations in multiple genes by CRISPR/Cas-mediated genome engineering.* Cell, 2013. 153(4): p. 910-8.

[89] Hockemeyer, D., et al., *Genetic engineering of human pluripotent cells using TALE nucleases.* Nat Biotechnol, 2011. 29(8): p. 731-4.

[90] Joung, J.K. and J.D. Sander, *TALENs: a widely applicable technology for targeted genome editing.* Nat Rev Mol Cell Biol, 2013. 14(1): p. 49-55.

[91] Mussolino, C. and T. Cathomen, *TALE nucleases: tailored genome engineering made easy.* Curr Opin Biotechnol, 2012. 23(5): p. 644-50.

[92] Wei, C., et al., TALEN or Cas9-rapid, efficient and specific choices for genome modifications. J Genet Genomics, 2013. 40(6): p. 281-9.

Chapter 4

Human Embryonic Stem Cell-Derived Primitive and Definitive Hematopoiesis

Bo Chen, Bin Mao, Shu Huang, Ya Zhou,
Kohichiro Tsuji and Feng Ma

Additional information is available at the end of the chapter

http://dx.doi.org/10.5772/58628

1. Introduction

It is well believed that human embryonic stem cells (hESCs [1]) and induced pluripotent stem cells (hiPSCs [2]) are of great potential use for tissue substitutes (for example, blood cells) and to cure various congenital disorders. In mammals, hematopoiesis has already been precisely described in murine system but not yet in human. Early development of hematopoietic system can be well defined by a series of waves from primitive hematopoiesis (early embryogenesis) to definitive ones (late fetal stages). In vitro induction of undifferentiated hESC to functionally mature blood cells may mimic the early hematopoietic development during human embryonic and fetal stages. It also provides an ideal model to uncover molecular and cellular mechanisms controlling early development of human hematopoiesis. On the other hand, functionally matured blood cells derived from hESC/hiPSCs are expected to be widely used for clinical cellular therapies. Although almost all kinds of the mature blood cells can be generated from hESCs, there still lacks solid evidence for the generation of reconstituting hematopoietic stem cells (HSCs) from hESC or hiPSC. So far until now, in vitro hESC-derived blood cells possess phenotypical maturity and partial functions while still more or less share embryonic/fetal characteristics, differing greatly from their adult counterparts. This indicated that in vitro culture systems are not perfect enough to exert full mature activities. Lack of knowledge about the molecular and cellular regulations in human early hematopoiesis has handicapped the development of research on hESC/hiPSC-derived hematopoiesis.

Having been focusing on basic and clinical research on hESC/hiPSC-derived functionally mature blood cells for long, our group has established an efficient method to induce large-scale production of multipotential hematopoietic progenitor cells by coculturing hESC/hiPSCs with murine hematopoietic niche-derived stromal cells [3-6]. By this method, large quantity of matured erythrocytes and other functional blood cells could be harvested. In this chapter we will discuss the latest progress in this research field along with our recent discoveries. We will emphasize on the origin, evolution and the development of both primitive and definitive hematopoietic waves, especially those derived from hESCs in vitro systems. The critical problems need to be solved and the research prospects of this field will also be addressed at the end of the chapter.

2. The primitive and definitive waves of hematopoiesis in mammals

2.1. Anatomical sites of hematopoiesis at different developmental stages

Hematopoiesis takes place in some discrete anatomical niches that change temporally and spatially in mammals. Its maturation along with developing ontogeny is a successive event initiated from yolk sac (YS) and then to intra-embryonic sites. The classical opinion believes that para-aortic splanchnopleura (P-Sp)/aorta-gonad-mesonephros (AGM) ought to be the sole location for the emergence of intra-embryonic hematopoiesis, where the earliest HSCs exist. During midgestation, these HSCs move to fetal liver (FL), a predominant hematopoietic place until birth [7, 8]. Hematopoietic precursors seed the bone marrow (BM) in late gestation, where maintaining the principal site of HSC activity lifelong. However, a recent discovery provided solid proof that before HSCs enter the circulation, the embryonic day (E) 10.5–11.5 mouse head is an unappreciated site for HSC emergence within the developing embryo independent to the AGM region [9].

Yolk Sac. The first wave of blood cell generation begins at embryonic day E7.0 at YS and is termed primitive hematopoiesis, producing large erythroblasts that express embryonic hemoglobins [10, 11]. The second wave, termed definitive hematopoiesis, produces smaller erythroblasts that express adult hemoglobins and various other blood cells [12]. Long-term repopulating HSCs (LTR-HSC) appear only in the second wave [13].

The murine YS is a bilayer organ composed of extra-embryonic mesoderm cells apposing to visceral endoderm cells. Its mesoderm layer produces the first blood cells within blood islands [14, 15]. Between E8.0 and E9.0, the outer layer's cells of the blood island in YS differentiate into endothelial cells and form a spindle shape while the vast majority of the inner cells gradually lose their intercellular attachments along with their differentiation into primitive erythroblasts [16].

The research of Yoder's group support that the YS not only acts as the sole site of primitive erythropoiesis but also possibly serves as the first source of definitive hematopoietic progenitors during embryonic development [15,17]. The result of BFU-E assay indicated that following the early wave of primitive erythropoiesis, definitive erythroid progenitors appear at 1-7 somite pairs (E8.25) and solely exist within the YS. After that the definitive erythroid (according to the CFU-E assay), mast cell and bipotential granulocyte/macrophage progenitors develop in the YS [18]. Another proof comes from Ncx1-/- embryo, which could not initiate a heartbeat on embryonic day E8.25 while development continued through E10. There is similar amount of primitive erythroid progenitors and definitive HPCs in YS of Ncx1-/- and WT mice through E9.5, while the P-Sp region in Ncx1-/- mice lacks primitive erythroblasts and definitive hematopoietic progenitor cells (HPCs) from E8.25 to E9.5 [17]. So it is reasonable to believe that primitive erythroblasts and nearly all definitive HPCs seeded the fetal liver after E9.5 are generated from the YS between E7.0-E9.5 and are re-distributed into the embryo proper via the systemic circulation [17]. The definitive hematopoiesis may originate from primitive hematopoiesis during embryonic development and migrating HSCs come from the murine YS, which seed the liver and initiate hematopoiesis on 10.0 day postcoitus (dpc). But whether YS cells isolated before day 11.0 dpc possess any long-term repopulating HSC activity remains controversy. Yoder et al proved that donor day 9.0 dpc YS cells could establish long-term hematopoietic system in conditioned newborns, but not in adult recipients [19]. When these early YS cells were co-cultured for 4 days with AGM-S3, an AGM region-derived stromal cell line, they obtained such capacity of reconstitution [20]. All the evidence so far agree that YS itself couldn't provide long-term repopulating HSC activity while the hematopoietic cells originate from YS could own this activity through proper "education" by definitive hematopoiesis niches or the stromal cell line derived from them [20, 21].

AGM. Following the onset of circulation at E8.5, hematopoietic progenitors rapidly move within the embryo. Determining the anatomical origins of definitive hematopoiesis is a complicated and controversial topic. It is now widely recognized that main source of definitive hematopoiesis originates from the AGM region [8]. Using irradiated adult mice as the recipients, Müller et al reported that LTR-HSC first appeared in the AGM region at 10 dpc and expanded in 11 dpc AGM region [13]. Medvinsky et al then demonstrated that, at day 10 in gestation, hematopoietic stem cells initiate autonomously and exclusively within the AGM region under in vitro organ culture condition [8]. All these findings suggested that the AGM region at 10 to 11 dpc provides a microenvironment suitable for generation of LTR-HSC. Xu et al obtained a stromal cell line derived from the AGM region of 10.5 dpc mouse embryo (AGM-S3) that could support the growth and proliferation of hematopoietic progenitor/stem cells from adult mouse bone marrow and human cord blood without additional cytokines [21]. The same research group found that no definitive hematopoiesis-derived colony-forming cells were generated from YS and P-Sp cells at 8.5 dpc before co-cultured with AGM-S3 cells. However after 4-day co-culture of 8.5 dpc YS and P-Sp cells with AGM-S3, spleen colony-forming cells and HSCs capable of reconstituting definitive hematopoiesis in adult mice simultaneously appeared [20]. It is proposed that precursors that had the potential to generate definitive HSCs appear in both extra-embryonic (YS) and intra-embryonic (P-Sp) region, the latter providing microenvironment to support the definitive hematopoiesis from both precursors.

Fetal liver and bone marrow. At 9 dpc, the liver rudiment begins to form an evagination of gut into the septum transversum. The liver does not generate hematopoietic cells de novo but is instead colonized at late E9 by hematopoietic cells generated in other tissues [22; 23]. The first erythroblasts are visible in the liver at 9 dpc. From 10 dpc onwards, the erythroid lineage begins to develop definitive characteristics. Myeloid CFU-Cs appears in the fetal liver at 9 dpc and macrophages and B cells are present at 10-11 dpc [24]. Although most of these differentiated cell types are found early in liver development, the more immature cell types, such as the CFU-S progenitor and the LTR-HSC [7, 25], can be detected only beginning at 11 dpc. Since the liver rudiment is colonized by exogenous blood cells [26]HSCs must arise elsewhere.

Previous study on mouse embryo supported that HSCs able to engraft adult mice were present in the liver beginning at E11–E12 [27-29]. These fetal liver–derived HSCs expressed CD34, c-kit, AA4.1, and Sca-1 surface markers, and were thought to migrate to the fetal BM after E15 [30], the latter providing a continuous supply of mature blood cells for the lifespan (Figure 1).

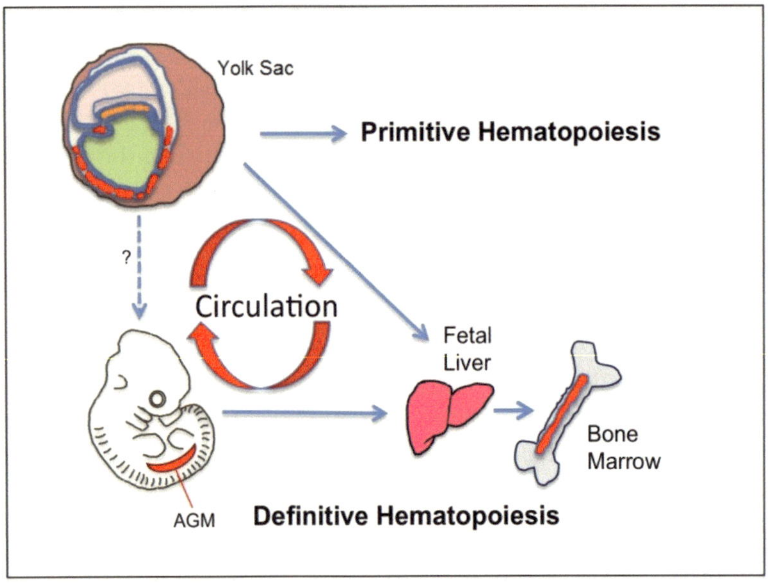

Figure 1. Primitive and definitive hematopoiesis

2.2. Characterization of the progenies derived from primitive and definitive hematopoiesis

The embryonic and fetal erythrocytes only come from primitive hematopoiesis wave while definitive HSCs and T lymphocytes only come from definitive hematopoiesis wave [31]. We will summarize the cellular and molecular characteristics of these cells in primitive and/or definitive hematopoiesis.

2.2.1. Erythrocytes

The erythroid cells within the blood islands of YS are known as primitive or embryonic erythrocytes, specially named as EryP. Differing from the ones found in the FL and adult BM, these YS primitive erythrocytes are typically large in size, nucleated and express embryonic pattren of hemoglobins [32, 33]. The development of primitive erythroid lineage is so transient that its progenitors reach top number in the developing YS at E8.25 and soon become undetectable at E9.0. They would never occur in the late stage of embryonic development [18, 33].

Following the early wave of primitive erythropoiesis and before the generation of adult-repopulating HSCs, a transient erythromyeloid wave of definitive hematopoietic progenitors (erythroid/myeloid progenitors [EMPs]) emerges in the YS. This YS-derived definitive wave begins E8.25 and colonizes the liver by E10.5. All maturation stage of erythroid precursors and the first definitive erythrocytes could be observed in the liver and circulation at E11.5 [34]. Initially, the development and differentiation of both primitive and definitive erythroid precursor cells depend on growth factor signaling [35],while the later committed erythroid progenitor is different [18, 36].

After HSC expansion in the fetal liver between E12.5 and E16.5, the definitive erythroid progenitors are exponentially increased, which produce massive number of definitive erythroid precursors exclusively expressing adult globins [37, 38]. Thus, HSCs that colonize the fetal liver are commonly described as the origin of "definitive" hematopoiesis in the mammalian embryo.

2.2.2. Macrophage

Another mature hematopoietic cells present in the early YS are macrophages. Progenitor cells of functional macrophages were first found in YS of the mouse between day 7 and 8 of gestation. Pro-monocytes and macrophages were firstly identified in the 10-day YS and 11-day FL, indicating that the primitive macrophage progenitors are "proximal" to the promonocyte on the pathway of sequential macrophage maturation [39]. The observation that they mature rapidly, bypassing the monocyte stage of development, and express lower level of certain genes than later stage macrophages suggests that they could represent a unique population [40, 41].

However, the distinction of primitive macrophages is not as clear as the primitive erythroid lineage, since it is unknown if the molecular events leading to their development differ significantly from other types of macrophages. Because that is a close relationship between microphages and the maturation of erythroid cells, the special developmental trait of microphages probably is the adaptation for the primitive erythrogenesis process [12, 18]

2.2.3. Mast cells

In fetal stages, a large number of MC precursors could be found in murine YS and fetal blood, indicating a strong wave of MC development taking place in early embryo [42, 43]. But little is known about human MC development during the embryonic and fetal stages. The human and non-human primates MCs could be generated from ES cells when co-cultured with mouse AGM or OP9 (a fetal bone cell line from M-CSF knockout mouse) cells [4, 44], providing important information about human MC development in early embryonic stage.

In humans, two types of MCs have been identified based on their neutral protease compositions [45]. Connective tissue-type MCs (CT-MCs)mainly located in skin and sub-mucosal area and express tryptase, chymase, MC carboxypeptidase, and cathepsin G in their secretory granules. Mucosal-type MCs (M-MCs) that locates in alveolar wall and small intestinal mucosa express only tryptase. The non-human primate ESC-derived MCs are similar to CT-MC in phenotype and functionally identical to human skin counterparts, indicating that a different pathway may occur in early development for this two type of MCs. [4].

2.2.4. *Lymphocytes*

It is ambiguous to obtain common opinion about the origin site and time of lymphocytes during embryonic development. Some researchers persist that lymphoid precursors could be detected in the YS (extraembryo) prior to the embryo, which is as early as Day 8 of gestation [46, 47]. The opposite opinion believe that lymphoid precursors appear in the embryo proper before the YS [48]. Other researchers believe that they appear in both sites at the same developmental stage [49, 50]. The clonal assays of Godin et al. suggested that this controversy might be due to difference in the experimental condition among these groups to favor lymphoid potential but not lymphoid commitment [51]. If the latter as a standard, the early fetal liver was the first place to produce committed lymphoid precursors [52]. Different to myelopoiesis, lymphogenesis has no primitive wave and is specifically derived from definitive hematopoiesis.

2.2.5. *LTR-HSC*

The definitive hematopoiesis is defined by LT-HSCs that could reconstitute the hematopoietic system in irradiated adult mouse. The primitive hematopoiesis could not support such a reconstitution in adult mouse but only in fetal ones [8]. The primitive hematopoiesis has too simple and incomplete hematopoietic hierarchy compared with definitive hematopoiesis and could not be detected of the complete activity of LT-HSC [24]. The mechanism controlling developmental difference between two waves of hematopoiesis has been obviously observed, but whether "primitive" HSC exists is still a controversial topic [53]. Some researches support the opinion that LT-HSC also originates from YS. Only after the "education" on AGM, these YS LT-HSC progenitors become the functional LT-HSCs, for the hematopoietic cell isolated from E9.5 YS could reconstitute hematopoietic system in irradiated adult mouse after 4 days co-cultured with mAGM-S3 cells [20]. Immediately after primitive hematopoiesis at YS, the definitive hematopoietic progenitors could be detected by generating CFU-GM and BFU-E, indicating that HSCs probably originate from YS and finally mature at AGM region [18]. But other groups persist that LT-HSC has no relation to YS and was produced de novo from AGM [54].

2.3. Molecular mechanisms controlling the transition from primitive to definitive hematopoiesis

Recently, findings from different gene targeting experiments have demonstrated that the primitive and definitive hematopoietic lineages develop from a common precursor by distinct molecular programs, and that the respective cell populations are regulated by different growth factors. Though the controversy still exists about their origins, the YS-derived cell population that only have primitive hematopoiesis potential could obtain definitive hematopoiesis potential by the "education" on mAGM cells in vitro [20]. The YS cell grafted to fetal liver or fetal marrow could also obtain LT-HSC activity [19, 55]. This provides strong evidences to support that the primitive hematopoietic progenitor cells could transfer to definitive ones in the microenvironment of fetus definitive hematopoiesis origin, such as mouse AGM, fetal liver and fetal marrow. It is obvious that molecular signals released by definitive hematopoiesis niche cells play a key role in such "education".

Because the component of hemoglobin provides clear trace to describe the developmental stages of red blood or its precursor that distinguish the primitive from definitive hematopoiesis, the erythropoiesis serves as an ideal model to research both waves of hematopoiesis. It also provides clue to dissect their molecular switch mechanism, which is also the main object in our discussion about the molecular regulation of both primitive and definitive hematopoiesis.

In situ studies of the early embryo have demonstrated that genes known to play a role in the onset of hematopoietic development (e. g., GATA-2,scl/tal-1, rbtn2) are expressed prior to the appearance of the blood islands [56]. This suggests that the molecular program that leads to hematopoietic commitment begins shortly after gastrulation at approximate Day 7.0 of gestation. Under the control of tal-1, rbtn2, GATA-2, and GATA-1 that was expressed orderly, primitive mammalian erythropoiesis takes place in a subpopulation of extra-embryonic mesoderm cells during gastrulation. Though these transcriptional factors were also expressed in other region of embryo in the same stage (tal-1 and rbtn2 also in posterior embryonic mesoderm and GATA-1 and GATA-2 expression also in extra-embryonic tissues of ectodermal origin), their expression pattern in extra-embryonic mesoderm cells still play a key role in understanding the molecular mechanism of hematopoietic commitments [56].

By gene-targeting studies on transcription factors essential for development of all hematopoietic lineages, the key control genes for primitive and/or definitive hematopoiesis (such as Gata1, Gata2, AML1, C-myb, EKLF, rbtn2) were well analyzed.

GATA gene family. GATA1 expression is highly restricted in erythroid cells, megakaryocytic cells, eosinophils, dendritic cells, and MCs of hematopoietic cell [57]. It is essential for red blood cell (RBC) development because GATA-1-/- mice will die between E10.5–E11.5 in mid-gestation by anemia [58]. In such mutated embryo, only primitive erythroid cells could be found in the peripheral blood, which are arrested at a proerythroblast stage and express βH1, α and ζ-globin transcripts, then die by apoptosis [59]. Matured definitive erythroid cells were completely absent in GATA-1-/- mice [60-62]. GATA-2, another important member of GATA family, was expressed in many other multi-lineage progenitors and HSCs [63] and also plays important role in proliferation, survival and differentiation of early hematopoietic cells, though the result of its functional mutation is relatively mild compared with GATA-1 [59].

At the late stage of erythrocyte development the "GATA switch" is the key molecular mechanism for erythroid differentiation companied by down-regulating GATA2 and up-regulating GATA1 expression. This is the process that GATA-1 occupies the GATA binding site on the upstream element of GATA2 gene and represses the expression of the latter [64]. In general, after terminal erythroid differentiation start, GATA1 directly opens the expression of erythroid lineage-affiliated genes such as β-globin, Alas2, and Gata1 itself while at the same time represses Gata2, c-Kit, c-Myb, and c-Myc, responding for the proliferation of progenitors in earlier stages of hematopoiesis [65].

AML1/Runx1. Homozygous mutations of AML1/Runx1 did not interfere normal morphogenesis and YS-derived erythropoiesis, but completely inhibit FL hematopoiesis, leading to the death of embryo around E12.5. The same mutation in ES cells do not influence their differentiation potential into primitive erythroid cells in vitro while stop occurence of any definitive myeloid or erythroid progenitors in both the YS and FL after injected to blastocysts to produce chimeric animals. Above proofs support the key role of AML1 and AML1-regulated target genes to all lineages in definitive hematopoiesis [10, 66]

c-Myb. c-Myb is highly expressed in immature hematopoietic cells and its expression is down-regulated as they become more differentiated [67, 68]. c-Myb controls self-renewal and differentiation of adult HSCs and its disruption seriously depletes the HSC pool and inhibits the definitive hematopoiesis [69].

EKLF. EKLF is zero or very weakly expressed in hematopoietic stem cells and multipotential myeloid progenitors, while arise since more matured stages and play a role all the time later for RBC differentiation. Its expression is up-regulated when myoloid and erythriod progenitors were committed to the erythroid lineage while down-regulated when differentiate toward megakaryopoiesis [70]. During the global expansion of erythroid gene expression in primitive and definitive lineages, EKLF also plays a direct role in globin switching. EKLF is weakly expressed during embryonic and fetal development, which led to a low expression of adult β-globin, Bcl11a and a high one of γ-globin. While in adults, EKLF is highly expressed in definitive RBCs that results in high levels of adult β-globin and Bcl11a expression, and represses γ-globin expression [71]. Finally, EKLF stop the cell cycle of RBCs at the terminal maturation [72]. EKLF/KLF1 mutations will change the RBC phenotypes or even lead to disease. [73, 74]

Rbtn2. Rbtn2 is a nuclear protein expressed in erythroid lineage in vivo, which is essential for erythroid development in mice. The homozygous mutation of rbtn2 inhibits YS erythropoiesis and leads to embryonic lethality around E10.5. YS tissue from homozygous mutant mice and double-mutant ES cells could not process erythroid development in in vitro differentiation system, showing a key role for Rbtn2 in erythroid differentiation, which is high related to GATA-1 [75]

EPO/EPOR. Erythropoietin is a glycoprotein produced primarily by kidney and is the principal factor to regulate RBC production, mainly functioning on erythroid progenitors within the FL and adult BM [76]. The erythroid progenitors before BFU-E stage and RBC after late basophilic erythroblast stage are not responsive to EPO. Proliferation in CFU-E stage could highly responsive to EPO and this response is very transient. The affinity between erythropoietin and its receptor (EPO-R) and their concentration decide strength of such response during the erythropoiesis. EPO-R signaling pathway is necessary for both primitive and definitive erythropoiesis [77, 78].

http://dx.doi.org/10.5772/58628

3. hESC-derived primitive and definitive hematopoietic cells

The first hESC line was established by Thomson's group in 1998 [1] and then the first hiPSC line by Yamanaka's group in 2007 [2]. Both cells provide possibility to uncover various normal or diseased mechanisms in early human development. By in vitro differentiation system, factors controlling the primitive and definitive hematopoiesis could be investigated in detail using the method of embryoid body (EB) forming [79], or co-culture with hematopoietic niche-derived stromal cell lines [3-6, 80, 81]. Among them, method of EBs could obtain large quantity of blood cells that were not well matured, mimicking the primitive hematopoietic cells [82, 83]. Then, co-culture with OP9 cells was applied to promote the differentiation based on EB method and large-scale production of mature erythrocytes with some definitive properties could be obtained [84, 85]. Although OP9 co-culture system could obtain robust growth of matured blood cells, it is clearly not a natural process and could not be used as proper model to elucidate the natural mechanism controlling early human hematopoiesis. The stromal cells isolated from early hematopoietic niches, such as AGM region, FL and late-stage fetal BM, should be more reasonable candidates to support the in vitro differentiation of hESC/hiPSC-derived hematopoietic stem/progenitor cells. The Lako's group compared the differences among several in vitro differentiation systems based on stromal cell co-culture and optimized their culture conditions. Their result suggested that AGM-derived cell line was most proper for the hematopoiesis differentiation of ESC. [86]

Based on in vitro differentiation systems, researchers tried to make clear the details of early human hematopoiesis using hESCs as a model. Keller's group found two distinct types of hemangioblasts during hESCs differentiation culture, one could give rise to primitive erythroid, macrophages and endothelial cells, while the other one generated only primitive erythroid and endothelial cells [87]. Their work provided the first evidence to prove the existence of hemangioblasts derived from hESCs in vitro. The follow-up work showed that under the control of growth factors and hematopoietic cytokines the hematopoiesis and myelopoiesis will happen during the later stage; and common bi-potent progenitor capable of generating erythroid and megakaryocytic cells could be observed, mimicking the process in vivo [88, 89]. Through the analysis of hemoglobin components, an erythroid maturation could be observed by the ratio of β-globin expression during hESC differentiation culture, which seems a maturation switch but not lineage switch [90]. Above researches indicated that the in vitro hESC differentiation system could reflect the in vivo hematopoietic process during the early human embryonic development.

Slukvin's group applied PO9 co-culture system to explore the detail pathway from hESCs to definitive hematopoiesis. They firstly characterized a population of differentiating hematopoietic cells defined by the expression of CD43, which is distinct to endothelial and mesenchymal cells. Then they defined the erythro-megakaryocytic progenitors (CD34+CD43+CD235a+CD41a+/-CD45-) and multi-potent lymphohematopoietic progenitors (CD34+CD43+CD235a-CD41a-CD45-) in later stage, which replicated the beginning of definitive hematopoiesis in some degree [91, 92]. The human myelomonocytic cells could also be generated from expansion and differentiation of pluripotent stem cell-derived lin-

CD34+CD43+CD45+ progenitors. All the defined population above could be detected during the in vitro differentiation for different hESCs or hiPSCs lines [92].

By different culture conditions, nearly all of the blood lineages could be obtained from hESCs according to the experience from murine ESC protocols, which pave the way to the clinic application [93]. Jame's group also identified a wave of hemogenic endothelial development during the transition from endothelial to hematopoietic cell [94], reflecting the classic property of definitive hematopoiesis. Their result showed that the definitive hematopoiesis could be researched in vitro differentiation system and most progenitor population could be obtained in such a system.

Although research on the early development of human hematopoiesis using human embryo is rigorously restricted by ethics, so far accumulated data have clearly demonstrated that in vitro hESC-derived hematopoiesis is more or less similar to the events happening in murine fetal development (Figure 2). Thus, research on hESC-hematopoiesis should contribute greatly to understand the developmental controls of human early hematopoiesis.

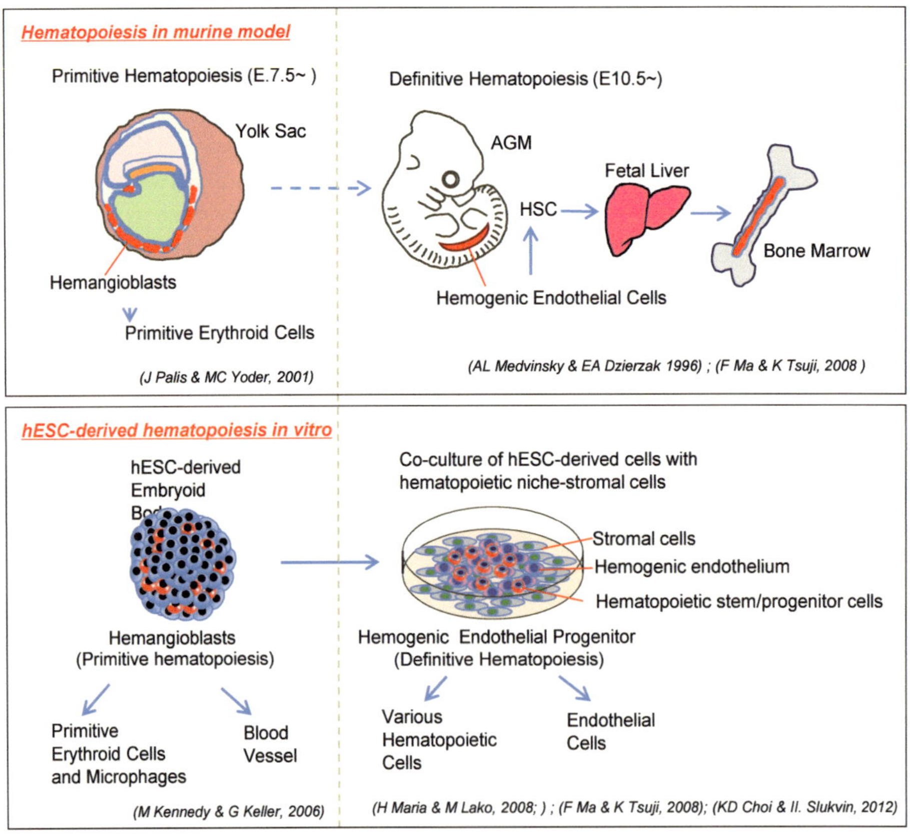

Figure 2. Comparison of mouse model and in vitro hESC/hiPSC-derived hematopoiesis

3.1. hESC-derived erythrocytes

Research on hESC-derived erythrocytes is one of the hot points because of the fact that development of erythrocytes shows distinct trait between primitive and definitive waves. More important is that the matured RBCs have no nucleus, which could avoid the exogenous gene interference when clinically applied to treat patients. Much labor has been donated to optimize the method of in vitro differentiation of RBCs from hESCs. For example, Lu's group firstly use EBs formation and then co-culture with OP9 to obtain a high yield of erythrocytes with some efficiency of enucleation. However, these erythrocytes were still immature because only 16% of these hESC-derived erythrocytes expressed β-globin [84]. Similar work has been done by several other groups using OP9 system [95, 96]

In our laboratory, we have recently established efficient blood cell-yeilding systems by co-culture of hESC/hiPSCs with murine AGM and FL stromal cells (Figure 3). In the co-culture, undifferentiated hESC/hiPSC colonies grow up and differentiate firstly to a mesoderm-like structure, then to hematopoietic progenitor cells on days 10 to 14. In the second suspension culture, these hESC/hiPSC-derived hematopoietic progenitors are further induced to some specific blood cell lineages, such as erythrocytes, mast cells, and eosinophils, etc. [3]. By a clone-tracing method, we gained concrete results that hESC-derived erythrocytes kept continuously progressing toward maturation over a time course. The expression of β-globin vs ε-globin showed a typical switching pattern, mimicking the normal development of human erythrocytes. On day 12 of the co-culture with murine FL-derived stromal cells, hESC-derived erythrocytes (BFU-E) express β-globin at about 60%, but up-regulated to almost 100% with additional 6 days of culture. These matured hESC-derived erythrocytes can undergo enucleation and release oxygen [5].

Our results showed that the in vitro differentiation from hESCs to matured hematopoietic cells is a progressive process if proper co-culture condition provided. Functionally matured blood cells similar to those from the definitive hematopoiesis could be obtained in large-scale production by this method.

3.2. hESC-derived HSCs

The most challenging task for research on hESC/hiPSC-derived hematopoiesis is to obtain the real HSC from in vitro differentiation system [97]. Although hESC/hiPSC-derived hematopoietic cells that could be engrafted in immune compromised mice have been reported [86, 98-102], effort to obtain real HSCs from hESCs has largely been done in vain during past years. In most experiments, the engraftment rate was very low and mostly restricted to the myeloid lineage, and it was ambiguous if these engrafed cells were derived from real HSCs. The co-culture with S17 could induce hESCs to HSC-like properties with low capacity of RBC potential [103]. If a modified cell line generated from mAGM (AM20.1B4) was used the RBC activity of HSC will be increased much [86]. But these hESC-derived HSC-like cells could not repopulate in NOD-SCID mice, representing they were not true definitive HSCs that satisfy the functional definition.

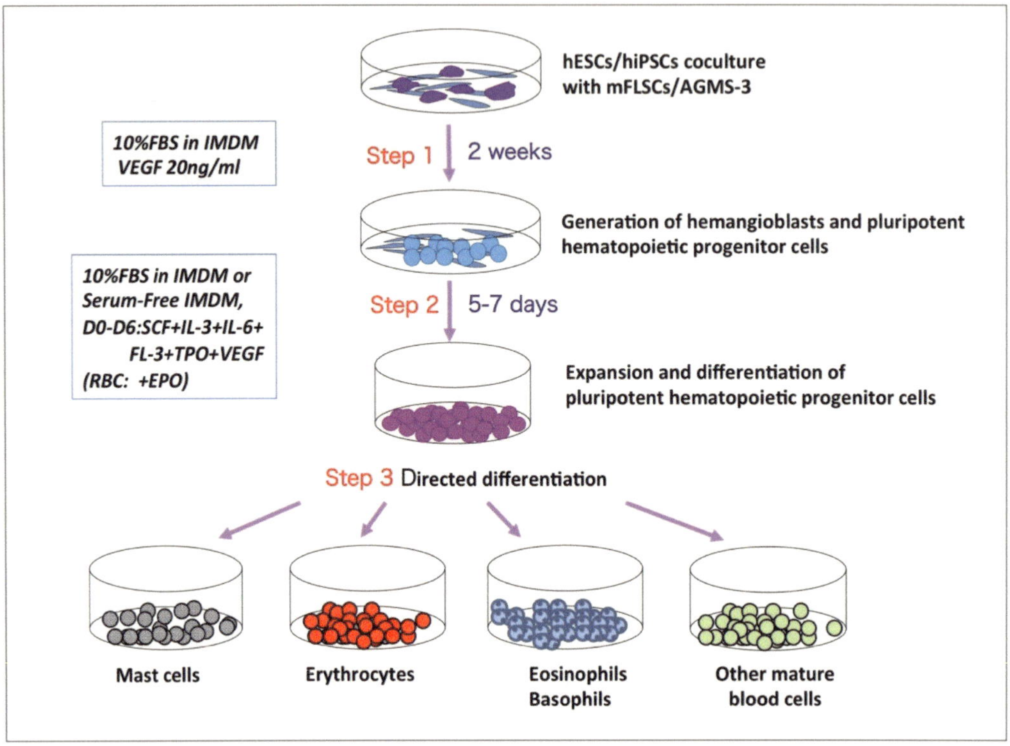

Figure 3. Co-culture of hESC/hiPSCs with murine AGM or FL stromal cells

3.3. hESC-derived lymphocytes

A recent report showed that when co-cultured with OP9-DL1 cells, the hESC-derived hematopoietic progenitor cells that formed endothelium-lined cell clumps could be induced into T-lineage cells [104]. These hESC-derived T cells expressed sequential surface markers from T-lymphoid progenitors (CD34+CD7+) to matured T cells (double positive CD4+CD8+and finally maturedCD3+CD1-CD27+). The T-lineage cell production provided concrete evidence that hESC-derived hematopoiesis endowed with definitive property. The functional matured hESC-derived NK cells could also be generated by sequential co-culture on different feeder cells [105, 106]. But the detail of HMC molecular and globulin class-switching during this process has still not been elucidated.

3.4. Hemangioblasts

The blood islands region in YS consists of clusters of primitive erythroblasts surrounded by mature endothelial cells. The close development relationship between hematopoietic and endothelial cells in such a region indicated that these lineages ought to share a common progenitor named as the hemangioblast [16, 33, 107]. This concept was provided nearly one century ago [108,109] while it has been circumstantially supported by the latest molecular genetic and embryological proofs [110-114]. More direct evidence came from the in vitro differentiation system based on ESCs. Some differentiation models identified a progenitor with properties of the hemangioblast [115, 116], indicating that the primitive hematopoiesis derived from ESCs should pass through a hemangioblast stage. Further studies showed that these hemangioblasts might have more potential than hematopoietic and endothelial cells, such as smooth muscle [117, 118].

hESC-derived hemangioblasts have also been identified in differentiation cultures [87]. These hESC-derived hemangioblasts were defined by the expression of KDR and could generate clonal cells sharing both hematopoietic and vascular potential. There were two distinct types of hESC-derived hemangioblasts: one gave rise to primitive erythroid cells, macrophages, and endothelial cells and the other generated only the primitive erythroid population and endothelial cells. This finding provided evidence that hESC-derived hematopoiesis mimicked the normal development pattern of human earliest stage of hematopoietic commitment.

3.5. Hemogenic endothelium

Different to hemangioblasts from YS, the concept of "hemogenic endothelial" come from the research of AGM, which is in the ventral wall of the aorta and buds off HSCs [13]. The molecular control of hemogenic endothelium is different to that of hemangioblasts. For example, Runx1 was indispensable for the hematopoiesis originated from hemogenic endothelia, but not hemangioblasts [119, 120]. And the hemogenic endothelium did not originate from hemangioblasts but presumptive mesoangioblasts, which could express endothelial-specific genes and ultimately express HSC-associated markers [121]. Since hemogenic endothelium is closely related to definitive hematopoiesis and has been regarded as the necessary stage for generation of HSCs, the details of its potential and developmental process to produce HSCs should be finely elucidated [122-125].

In some in vitro differentiation systems, hESC-derived CD34+ hemogenic progenitor cells could also be detected with the endothelium potential [3, 126]. In an OP-9 co-culture system, hESC-derived hemogenic endothelium progenitors (HEPs) were identified pinpoint by VE-cadherin+CD73-CD235a/CD43- intermediate phenotype, which arise at the post primitive streak stage of differentiation directly from a hematovascular mesodermal precursor (KDR +APLNR+PDGFRa$^{low/-}$). These HEPs differ from non-HEPs (VE-cadherin+CD73+) and early hematopoietic cells (VEcadherin+CD235a+CD41a−) [124]. This subtle finding may provide clue to facilitate generation of HSCs from hESCs (Figure 4).

3.6. Switching mechanisms controlling primitive hematopoiesis to definitive one

The origin of two waves of hematopoiesis was distinct in the sight of embryogenesis by the research for Xenopus [127], mice [128] and human [54]. However, the reciprocal transplant test in Xenopus embryo proved that the hematopoietic progenitor cell in VBI(corresponding to YS) and DLP(corresponding to AGM) could change their potential according to microenvironment of the graft site [129]. Similarly, blast-like cells derived from murine ESCs could differentiate to both primitive and definitive lineages [130]. In our study by a clonal tracing method, hESC-derived erythrocytes showed the primitive properties at the early stage and progressively

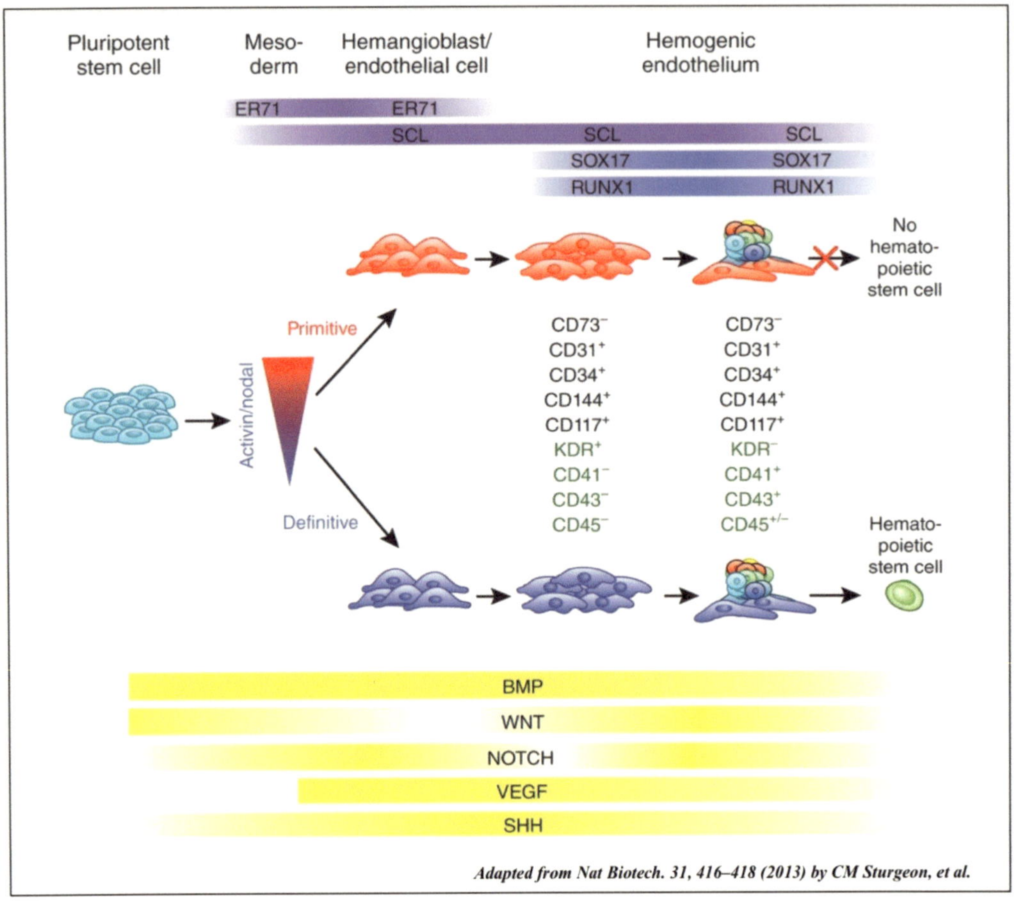

Figure 4. Development of hemangioblasts and hemogenic endothelia from pluripotent stem cells

turned to be definitive ones along with co-culture on FL cells [5], indicating that there exists a possible switching mechanism controlling the conversion.

However, the molecular mechanism controlling such a switch remains unclear so far. The switch observed in process of hESC-derived hematopoiesis probably is caused by the influence or induction of the definitive hematopoietic niche-derived stromal cells (AGM, FL). The cellular and molecular signals, including the growth factors and adhesion proteins expressed by feeder cells and growth factors added in medium ought to play a key role to activate the switch controlling the definitive hematopoiesis (such as Hedgehog, Notch1 and BMP signal pathways). Among them, HoxB4 had been identified as the key factor to promote such a switch [131]. Many key factors expressed in primitive and/or definitive erythrogenesis could also be tracked in ESC-derived hematopoiesis [94, 132]. Since the hemoglobin switch of erythrocyte could be used as the indicator of premitive and definitive hematopoiesis, the erythrogenesis-related factors, such as GATA1, GATA2, AML1, SCL, EKLF, ought to be changed their expression profile during such a switch. A thorough transcriptome analysis by deep sequencing or gene chip should be useful tools to detect hESC-derived hematopoietic switch.

4. Concluding remarks and future perspectives

The in vitro production of functionally matured blood cells from hESC/hiPSCs has provided an excellent model for both basic research and clinical applications. hESC/hiPSC-derived RBCs are highly fancied because of their unlimited use as the substitute for blood transfusion, not to mention the the utilization of hESC/hiPSC-derived HSCs in transplantation. On the other hand, hiPSCs can provide patient-tailored models for analyzing the pathogenesis of malignant blood diseases and thus develop individual treatment by molecular corrections. Innate immune-related cells differentiated from hESC/hiPSCs will also help us to reveal the initial establishment of human immune system, and to explore drug screening system and cell therapy model for deficiencies of innate immunity.

However, before the successful application of hESC-derived cellular therapy, there are still many problems needed to be solved. The molecular mechanisms controlling both primitive and definitive hematopoiesis should be clarified at first. The HSC-like cells and functional matured blood cells derived from hESC / hiPSC in vitro must also be proven their activities in vivo. Besides, more efficient culture system free from xenobiotics must be optimized.

In order to realize above aims, the following efforts need to be done to promote the research in this field:

1. The latest molecular biological technology and other new technique tool should be applied in the future research.

Currently, Tetracycline (tc) and Tet repressor (TetR) system (tet-on and off) is one of the most matured and widely applied regulatory systems [133] in organisms ranging from bacteria to mammals. The technique by Transcription activator-like effector nucleases (TALENs [135, 136]) and clustered regularly interspaced short palindromic repeats (CRISPR)/CRISPR-associated (Cas) systems [136, 137] have also been developed. These methods let us be able to do molecular manipulation in genomic scale(named "genome edit") by unlimited times against single cell line without obvious influence to the cell characteristic, thus provide an ideal model to investigate the development events in hESC-derived hematopoiesis.

2. A highly efficient animal-source-free 3D culture system should be developed to search for the generation of hESC-derived HSCs.

A big problem that hampers the progress in research on hESC-derived hematopoiesis is a lack of a proper culture system that fully mimics the microenvironment of early development of human hematopoiesis. An ideal culture model should be completely free of any animal source substitutes and to great extent imitate a live structure as done in vivo circumstances. Since HSCs develop in a complicated niche composed of various cell types (endothelial cells, osteoblasts, mesenchymal cells, etc.), a 3D culture model mixed by several hematopoietic niche cells should benefit the efficiency of generating true HSCs from hESCs.

3. The new concept and theory should be introduced to conduct the research of hESC-derived hematopoiesis.

According to the recent researches by Deng and other group [138, 139], the pluripotency of stem cells is only the balance between several transcription factor groups controlling layer lineage differentiation. Their finding denied the existence of special pluripotent factors and showed that maintenance of ESC/iPSCs may be only because that all the way to any differentiation direction has been blocked by mutually antagonistic lineage specifiers [140]. Such concept may also help us to uncover the controlling balance by some specifiers in early hematopoiesis, especially for those controlling primitive and definitive ones. Since we always try to obtain more definitive blood cells for clinic application, to discover the key factor leading to definitive hematopoiesis is a challenging task to do. Moreover, how to define and describe the essence of HSCs at molecular level is also an issue need to be re-addressed.

Author details

Bo Chen[1], Bin Mao[1], Shu Huang[1], Ya Zhou[1], Kohichiro Tsuji[1] and Feng Ma[1,2*]

*Address all correspondence to: mafeng@hotmail.co.jp

1 Institute of Blood Transfusion, Chinese Academy of Medical Sciences & Peking Union Medical College, Chengdu, China

2 Division of Stem Cell Processing, Center for Stem Cell Biology and Regenerative Medicine, Institute of Medical Science, University of Tokyo, Tokyo, Japan

References

[1] Thomson JA, Itskovitz-Eldor J, Shapiro SS, Waknitz MA, Swiergiel JJ, Marshall VS, Jones JM, Embryonic stem cell lines derived from human blastocysts. *Science* 1998; 282 (5391): 1145-7.

[2] Takahashi K, Tanabe K, Ohnuki M, Narita M, Ichisaka T, Tomoda K, Yamanaka S,. Induction of Pluripotent Stem Cells from Adult Human Fibroblasts by Defined Factors. *Cell* 2007;131 (5): 861-72.

[3] Ma F, Wang D, Hanada S, Ebihara Y, Kawasaki H, Zaike Y, Heike T, Nakahata T, Tsuji K. Novel method for efficient production of multipotential hematopoietic progenitors from human embryonic stem cells. *Int. J. Hematol*. 2007; 85 (5): 371-9.

[4] Ma F, Kambe N, Wang D, Shinoda G, Fujino H, Umeda K, Fujisawa A, Ma L, Suemori H, Nakatsuji N, Miyachi Y, Torii R, Tsuji K, Heike T, Nakahata T. Direct devel-

opment mature tryptase/chymase double positive connective tissue-type mast cells from primate ES cells. *StemCells* 2008; 26 (3): 706-714

[5] Ma F, Ebihara Y, Umeda K, Sakai H, Hanada S, Zhang H, Zaike Y, Tsuchida E, Nakahata T, Nakauchi H, Tsuji K. Generation of functional erythrocytes from human embryonic stem cell-derived definitive hematopoiesis. *roc. Natl. Acad. Sci. USA* 2008; 105 (35): 13087-92.

[6] Ma F, Yang WY, Ebihara Y, Tsuji K. Generation of Blood Cells from Human Embryonic Stem Cells and Their Possible Clinical Utilization. In:*Embryonic Stem Cells-Recent Advances in Pluripotent Stem Cell-Based Regenerative Medicine*. Craig Atwood: InTech; 2011:239-250.

[7] Medvinsky AL, Samoylina NL, Müller AM, Dzierzak EA. An early pre-liver intraembryonic source of CFU-S in the developing mouse. *Nature* 1993; 364 (6432): 64-67.

[8] Medvinsky AL, Dzierzak EA. Definitive hematopoiesis is autonomously initiated by the AGM Region. *Cell* 1996; 86 (6):897-906.

[9] Li Z, Lan Y, He W, Chen D, Wang J, Zhou F, Wang Y, Sun H, Chen X, Xu C, Li S, Pang Y, Zhang G, Yang L, Zhu L, Fan M, Shang A, Ju Z, Luo L, Ding Y, Guo W, Yuan W, Yang X, Liu B. Mouse embryonic head as a site for hematopoietic stem cell development. *Cell Stem Cell* 2012; 11 (5):663-75.

[10] Okuda T, van Deursen J, Hiebert SW, Grosveld G, Downing JR. AML1, the target of multiple chromosomal translocations in human leukemia, is essential for normal fetal liver hematopoiesis. *Cell* 1996; 84 (2): 321-30.

[11] Ferkowicz MJ, Starr M, Xie X, Li W, Johnson SA, Shelley WC, Morrison PR, Yoder MC. CD41 expression defines the onset of primitive and definitive hematopoiesis in the murine embryo. *Development* 2003;130 (18):4393-4403.

[12] McGrath KE, Palis J. Hematopoiesis in the yolk sac: more than meets the eye. *Exp. Hematol.* 2005; 33 (9):1021-1028.

[13] Müller AM, Medvinsky A, Strouboulis J, Grosveld F, Dzierzak E,Development of hematopoietic stem cell activity in the mouse embryo. *Immunity* 1994; 1 (4):291-301.

[14] Jollie, WP. Development, morphology, and function of the yolk-sac placenta of laboratory rodents. *Teratology* 1990; 41 (4): 361-81.

[15] Palis J, Yoder MC. Yolk-sac hematopoiesis: the first blood cells of mouse and man. *Exp. Hematol.* 2001; 29 (8):927-36.

[16] Haar JL, Ackerman GA. A phase and electron microscopic study of vasculogenesis and erythropoiesis in the yolk sac of the mouse. *Anat. Rec.* 1971; 170 (2): 199-223.

[17] Lux CT, Yoshimoto M, McGrath K, Conway SJ, Palis J, Yoder MC. All primitive and definitive hematopoietic progenitor cells emerging before E10 in the mouse embryo are products of the yolk sac. *Blood* 2008; 111 (7):3435-3438

[18] Palis J, Robertson S, Kennedy M, Wall C, Keller G. Development of erythroid and myeloid progenitors in the yolk sac and embryo properof the mouse. *Development* 1999; 126: 5073–5084.

[19] Yoder MC, Hiatt K, Mukherjee P. In vivo repopulating hematopoietic stem cells are present in the murine yolk sac at day 9. 0 postcoitus. *Proc. Natl. Acad. Sci. USA* 1997; 94 (13): 6776-6780.

[20] Matsuoka S, Tsuji K, Hisakawa H, Xu Mj, Ebihara Y, Ishii T, Sugiyama D, Manabe A, Tanaka R, Ikeda Y, Asano S, Nakahata T. Generation of definitive hematopoietic stem cells from murine early yolk sac and paraaortic splanchnopleures by aorta-gonad-mesonephros region-derived stromal cells. *Blood* 2001; 98 (1): 6-12.

[21] Xu MJ, Tsuji K, Ueda T, Mukouyama YS, Hara T, Yang FC, Ebihara Y, Matsuoka S, Manabe A, Kikuchi A, Ito M, Miyajima A, Nakahata T,. Stimulation of mouse and human primitive hematopoiesis by murine embryonic aorta-gonad-mesonephros-derived stromal cell lines. *Blood* 1998; 92 (6):2032-2040.

[22] Johnson GR, Moore MA. Role of stem cell migration in initiation of mouse fetal liver haemopoiesis. *Nature* 1975; 258:726–728

[23] Houssaint E. Differentiation of the mouse hepatic primordium. II. Extrinsic origin of the haemopoietic cell line. *Cell Differentiation* 1981; 10: 243

[24] Dzierzak E, Medvinsky A. Mouse embryonic hematopoiesis. *Trends Genet.* 1995; 11 (9):359-66.

[25] Godin IE, Garcia-Porrero JA, Coutinho A, Dieterlen-Lièvre F, Marcos MA. Para-aortic splanchnopleura from early mouse embryos contains B1a cell progenitors. *Nature* 1993; 364 (6432): 67-70.

[26] Faloon P, Arentson E, Kazarov A, Deng CX, Porcher C, Orkin S, Choi K,Basic fibroblast growth factor positively regulates hematopoietic development. *Development* 2000; 127 (9):1931-1941.

[27] Belaoussoff M, Farrington SM, Baron MH. Hematopoietic induction and respecification of A-P identity by visceral endoderm signaling in the mouse embryo. *Development* 1998; 125 (24):5009-5018.

[28] Pardanaud L, Dieterlen-Lièvre F. Manipulation of the angiopoietic/hemangiopoietic commitment in the avian embryo. *Development* 1999; 126 (4):617-627.

[29] Dyer MA, Farrington SM, Mohn D, Munday JR, Baron MH. Indian hedgehog activates hematopoiesis and vasculogenesis and can respecify prospective neurectodermal cell fate in the mouse embryo. *Development* 2001; 128 (10):1717-1730.

[30] Byrd N, Becker S, Maye P, Narasimhaiah R, St-Jacques B, Zhang X. Hedgehog is required for murine yolk sac angiogenesis. *Development* 2002;129:361–372

[31] Keller G, Lacaud G, Robertson S. Development of the hematopoietic system in the mouse. *Exp. Hematol.* 1999; 27 (5):777-87.

[32] Barker JE. Development of the mouse hematopoietic system. I. Types of hemoglobin produced in embryonic yolk sac and liver. *Dev. Biol.* 1968; 18 (1): 14-29.

[33] Brotherton TW, Chui DH, Gauldie J, Patterson M. Hemoglobin ontogeny during normal mouse fetal development. *Proc. Natl. Acad. Sci. USA* 1979; 76 (6): 2853-2857.

[34] McGrath KE, Frame JM, Fromm GJ, Koniski AD, Kingsley PD, Little J, Bulger M, Palis J. A transient definitive erythroid lineage with unique regulation of the beta-globin locus in the mammalian embryo. *Blood* 2011; 117 (17): 4600-4608.

[35] Chui DHK, Liao S-K, Walker K. Foetal erythropoiesis in Steel mutant mice. III. Defect in dfferentiation from BFU-E to CFU-E during early development. *Blood* 1978; 51:539–547.

[36] Wong PMC, Chung S-H, Reicheld SM, Chui DHK. Hemoglobin switching during murine embryonic development: evidence for two populations of embryonic erythropoietic progenitor cells. *Blood* 1986; 67:716–721.

[37] Kurata H, Mancini GC, Alespeiti G, Migliaccio AR, Migliaccio G. Stem cell factor induces proliferation and differentiation of fetal progenitor cells in the mouse. *Br. J. Haematol.* 1998; 101 (4):676–687.

[38] Marks PA, Rifkind RA. Protein synthesis: its control in erythropoiesis. *Science* 1972; 175 (25):955–961.

[39] Cline MJ, Moore MA. Embryonic origin of the mouse macrophage. *Blood* 1972; 39 (6): 842-849.

[40] Morioka Y, Naito M, Sato T, Takahashi K. Immunophenotypic and ultrastructural heterogeneity of macrophage differentiation in bone marrow and fetal hematopoiesis of mouse in vitro and in vivo. *J Leukoc Biol.* 1994; 55 (5):642-651.

[41] Faust N, Huber MC, Sippel AE, Bonifer C. Different macrophage populations develop from embryonic/fetal and adult hematopoietic tissues. *Exp. Hematol.* 1997; 25 (5): 432-444.

[42] Sonoda T, Hayashi C, Kitamura Y. Presence of mast cell precursors in the yolk sac of mice. *Dev. Biol.* 1983; 97:89-94.

[43] Rodewald HR, Dessing M, Dvorak AM, Galli SJ. Identification of a committed precursor for the mast cell lineage. *Science* 1996; 271:818–822.

[44] Kovarova M, Latour AM, Chason KD, Tilley SL, Koller BH. Human embryonic stem cells: a source of mast cells for the study of allergic and inflammatory diseases. *Blood* 2010; 115 (18): 3695-3703.

[45] Nakahata T, Toru H. Cytokines regulate development of human mast cells from hematopoietic progenitors. *Int J Hematol.* 2002; 75:350-356.

[46] Liu C, Auerbach R. In vitro development of murine T cells from prethymic and preliver embryonic yolk sac hematopoietic stem cells. *Development* 1991; 113:1315

[47] Palacios R, Imhof B. At day 8-8.5 of mouse development the yolk sac, not the embryo proper, has lymphoid precursor potential in vivo and in vitro. *Proc. Natl. Acad. Sci. USA* 1993; 90: 6581

[48] Ogawa M, Nishikawa S, Ikuta K, Yamamura F, Naito M, Takahishi K, Nishikawa SI. B cell ontogeny in murine embryo studies by a culture system with the monolayer of a stromal cell clones ST2: B cell progenitor develops first in the embryonal body rather than in the yolk sac. *EMBO J.* 1988; 7:1337

[49] Cumano A, Furlonger C, Paige C. Differentiation and characterization of B-cell precursor detected in the yolk sac and embryo body of embryos beginning at the 10-to 12-somite stage. *Proc. Natl. Acad. Sci. USA* 1993; 90:6429.

[50] Huang H, Zettergren L, Auerbach R. In vitro differentiation of B cells and myeloid cells from the early mouse embryo and its extraembryonic yolk sac. *Exp. Hematol.* 1994; 22:19

[51] Godin IE, Dieterlen-Lievre F, Cumano A. Emergence of multipotent hemopoietic cells in the yolk sac and paraaortic splanchnopleura in mouse embryos, beginning at 8.5 days postcoitus. *Proc. Natl. Acad. Sci. USA* 1995; 92:773.

[52] Lacaud G, Carlsson L, Keller G. Identification of a fetal hematopoieticprecursor with B cell, T cell and macrophage potential. *Immunity* 1998; 9:827.

[53] Costa G, Kouskoff V, Lacaud G. Origin of blood cells and HSC production in the embryo. *Trends Immunol* 2012; 33 (5): 215-223.

[54] Tavian M, Biasch K, Sinka L, Vallet J, Péault B. Embryonic origin of human hematopoiesis. Int. *J. Dev. Biol.* 2010; 54 (6-7): 1061-1065.

[55] Yoder MC, Hiatt K. Engraftment of embryonic hematopoietic cells in conditioned newborn recipients. *Blood* 1997; 89 (6): 2176-2183

[56] Silver L, Palis J. Initiation of murine embryonic erythropoiesis: a spatial analysis. *Blood* 1997; 89 (4):1154-1164.

[57] Leonard M, Brice M, Engel JD, Papayannopoulou T. Dynamics of GATA transcription factor expression during erythroid differentiation. *Blood* 1993; 82:1071–1079

[58] Fujiwara Y, Browne CP, Cunniff K, Goff SC, Orkin SH. Arrested development of embryonic red cell precursors in mouse embryos lacking transcription factor GATA-1. *Proc. Natl. Acad. Sci. USA* 1996; 93 (22): 12355-12358.

[59] Gregory T, Yu C, Ma A, Orkin SH, Blobel GA, Weiss MJ. GATA-1 and erythropoietin cooperate to promote erythroid cell survival by regulating bcl-xL expression. *Blood* 1999; 94:87–96.

[60] Pevny L, Simon MC, Robertson E, Klein WH, Tsai SF, D'Agati V, Orkin SH, Costantini F. Erythroid differentiation in chimaeric mice blocked by a targeted mutation in the gene for transcription factor GATA-1. *Nature* 1991; 349:257–260.

[61] Pevny L, Lin CS, D'Agati V, Simon MC, Orkin SH, Costantini F. Development of hematopoietic cells lacking transcription factor GATA-1. *Development* 1995; 121:163–172.

[62] Weiss MJ, Keller G, Orkin SH. Novel insights into erythroid development revealed through in vitro differentiation of GATA-1 embryonic stem cells. *Genes Dev.* 1994; 8:1184–1197.

[63] Suzuki N, Ohneda O, Minegishi N, Nishikawa M, Ohta T, Takahashi S, Engel JD, Yamamoto M. Combinatorial Gata2 and Sca1 expression defines hematopoietic stem cells in the bone marrow niche. *Proc. Natl. Acad. Sci. USA* 2006; 103:2202–2207.

[64] Suzuki N, Suwabe N, Ohneda O, Obara N, Imagawa S, Pan X, Motohashi H, Yamamoto M. Identification and characterization of 2 types of erythroid progenitors that express GATA-1 at distinct levels. *Blood* 2003; 102: 3575–3583

[65] Ferreira R, Ohneda K, Yamamoto M, Philipsen S. GATA1 function, a paradigm for transcription factors in hematopoiesis. *Mol. Cell. Biol.* 2005; 25:1215-1227

[66] Wang Q, Stacy T, Binder M, Marin-Padilla M, Sharpe AH, Speck NA. Disruption of the Cbfa2 gene causes necrosis and hemorrhaging in the central nervous system and blocks definitive hematopoiesis. *Proc. Natl. Acad. Sci. USA* 1996; 93 (8): 3444-9.

[67] Westin EH, Gallo RC, Arya SK, Eva A, Souza LM, Baluda MA, Aaronson SA, Wong-Staal F. Differential expression of the amv gene in human hematopoietic cells. *Proc. Natl. Acad. Sci. USA* 1982; 79 (7): 2194-2198.

[68] Gonda, TJ, Metcalf D. Expression of myb, myc and fos proto-oncogenes during the differentiation of a murine myeloid leukaemia. *Nature* 1984; 310 (5974): 249-251.

[69] Lieu YK, Reddy EP. Conditional c-myb knockout in adult hematopoietic stem cells leads to loss of self-renewal due to impaired proliferation and accelerated differentiation. *Proc. Natl. Acad. Sci. USA* 2009; 106 (51):21689-21694.

[70] Frontelo P, Manwani D, Galdass M, Karsunky H, Lohmann F, Gallagher PG, Bieker JJ. Novel role for EKLF in megakaryocyte lineage commitment. *Blood* 2007; 110 (12): 3871-3880.

[71] Bieker JJ. Putting a finger on the switch. *Nat. Genet.* 2010; 42 (9):733-734.

[72] Tallack MR, Keys JR, Perkins AC. Erythroid Kruppel-like factor regulates the G1 cyclin dependent kinase inhibitor p18INK4c. *J. Mol. Biol.* 2007; 369 (2):313-321.

[73] Siatecka M, Bieker JJ. The multifunctional role of EKLF/KLF1 during erythropoiesis. *Blood* 2011; 118 (8):2044-2054.

[74] Arnaud L, Saison C, Helias V, Lucien N, Steschenko D, Giarratana MC, Prehu C, Foliguet B, Montout L, de Brevern AG, Francina A, Ripoche P, Fenneteau O, Da Costa L, Peyrard T, Coghlan G, Illum N, Birgens H, Tamary H, Iolascon A, Delaunay J, Tchernia G, Cartron JP. A dominant mutation in the gene encoding the erythroid transcription factor KLF1 causes a congenital dyserythropoietic anemia. *Am. J. Hum. Genet.* 2010; 87 (5): 721-727.

[75] Warren AJ, Colledge WH, Carlton MBL, Evans M, Smith AJH, Rabbitts TH. The oncogenic cysteine-rich LIM domain protein is essential for erythroid development. *Cell* 1994; 78:45.

[76] Krantz SB. Erythropoietin. *Blood* 1991; 77 (3): 419-434.

[77] Longmore GD, Watowich SS, Hilton DJ, Lodish HF. The erythropoietin receptor: its role in hematopoiesis and myeloproliferative diseases. *J. Cell Biol.* 1993; 123 (6): 1305-1308.

[78] Richmond TD, Chohan M, Barber DL. Turning cells red: signal transduction mediated by erythropoietin. *Trends Cell Biol.* 2005; 15: 146-155.

[79] Keller G, Kennedy M, Papayannopoulou T, Wiles MV. Hematopoietic commitment during embryonic stem cell differentiation in culture. *Mol. Cell. Biol.* 1993; 13 (1): 473-486.

[80] Nakano T, Kodama H, Honjo T. Generation of lymphohematopoietic cells from embryonic stem cells in culture. *Science,* 1994; 265 (5175): 1098-1101.

[81] Palacios R, Golunski E and Samaridis J. In vitro generation of hematopoietic stem cells from an embryonic stem cell line. *Proc. Natl. Acad. Sci. USA* 1995; 92 (16): 7530-7534.

[82] Chang KH, Nelson AM, Cao H, Wang L, Nakamoto B, Ware CB, Papayannopoulou T. Definitive-like erythroid cells derived from human embryonic stem cells coexpress high levels of embryonic and fetal globins with little or no adult globin. *Blood* 2006; 108 (5):1515-1523.

[83] Chang KH, Huang A, Hirata RK, Wang PR, Russell DW, Papayannopoulou T. Globin phenotype of erythroid cells derived from human induced pluripotent stem cells. *Blood* 2010; 115 (12):2553-2554

[84] Lu SJ, Feng Q, Park JS, Vida L, Lee BS, Strausbauch M, Wettstein PJ, Honig GR, Lanza R. Biologic properties and enucleation of red blood cells from human embryonic stem cells. *Blood* 2008; 112 (12): 4475-4484.

[85] Ji J, Vijayaragavan K, Bosse M, Menendez P, Weisel K, Bhatia M. OP9 stroma augments survival of hematopoietic precursors and progenitors during hematopoietic differentiation from human embryonic stem cells. *Stem Cells* 2008; 26 (10): 2485-2495.

[86] Ledran MH, Krassowska A, Armstrong L, Dimmick I, Renström J, Lang R, Yung S, Santibanez-Coref M, Dzierzak E, Stojkovic M, Oostendorp RA, Forrester L, Lako M. Efficient hematopoietic differentiation of human embryonic stem cells on stromal cells derived from hematopoietic niches. *Cell Stem Cell* 2008; 3 (1): 85-98

[87] Kennedy M, D'Souza SL, Lynch-Kattman M, Schwantz S, Keller G. Development of the hemangioblast defines the onset of hematopoiesis in human ES cell differentiation cultures. *Blood* 2007; 109 (7): 2679-87.

[88] Grigoriadis AE, Kennedy M, Bozec A, Brunton F, Stenbeck G, Park IH, Wagner EF, Keller GM. Directed differentiation of hematopoietic precursors and functional osteoclasts from human ES and iPS cells. *Blood* 2010; 115 (14): 2769-76.

[89] Klimchenko O, Mori M, Distefano A, Langlois T, Larbret F, Lecluse Y, Feraud O, Vainchenker W, Norol F, Debili N. A common bipotent progenitor generates the erythroid and megakaryocyte lineages in embryonic stem cell-derived primitive hematopoiesis. *Blood* 2009; 114 (8): 1506-17.

[90] Qiu C, Olivier EN, Velho M, Bouhassira EE, 2008. Globin switches in yolk sac-like primitive and fetal-like definitive red blood cells produced from human embryonic stem cells. *Blood* 2008;111 (4): 2400-2408.

[91] Vodyanik MA, Thomson JA, Slukvin II. Leukosialin (CD43) defines hematopoietic progenitors in human embryonic stem cell differentiation cultures. *Blood* 2006; 108 (6):2095-105

[92] Choi KD, Vodyanik MA, Slukvin II. Generation of mature human myelomonocytic cells through expansion and differentiation of pluripotent stem cell-derived lin-CD34+CD43+CD45+progenitors. *J. Clin. Invest.* 2009; 119 (9):2818-2829.

[93] Olsen AL, Stachura DL, Weiss MJ. Designer blood: creating hematopoietic lineages from embryonic stem cells. *Blood* 2006; 107 (4): 1265-75.

[94] Rafii S, Kloss CC, Butler JM, Ginsberg M, Gars E, Lis R, Zhan Q, Josipovic P, Ding BS, Xiang J, Elemento O, Zaninovic N, Rosenwaks Z, Sadelain M, Rafii JA, James D. Human ESC-derived hemogenic endothelial cells undergo distinct waves of endothelial to hematopoietic transition. *Blood* 2013; 121 (5):770-780.

[95] Dias J, Gumenyuk M, Kang H, Vodyanik M, Yu J, Thomson JA, Slukvin II. Red blood cell generation from human induced pluripotent stem cells: perspectives for transfusion medicine. *Haematologica* 2010; 95 (10): 1651-1659.

[96] Lapillonne H, Kobari L, Mazurier C, Tropel P, Giarratana MC, Zanella-Cleon I, Kiger L, Wattenhofer-Donzé M, Puccio H, Hebert N, Francina A, Andreu G, Viville S,

Douay L. Red blood cell generation from human induced pluripotent stem cells: perspectives for transfusion medicine. *Haematologica* 2010; 95 (10):1651-1659.

[97] Slukvin II. Deciphering the hierarchy of angiohematopoietic progenitors from human pluripotent stem cells. *Cell Cycle* 2013; 12 (5): 720-727.

[98] Kaufman DS, Woll PHC, Martin JL, Xinghui TL. CD34+Cells derived from human embryonic stem cells demonstrate hematopoietic stem cell potential in vitro and in vivo. *Blood* 2004;104:163a

[99] Wang L, Menendez P, Shojaei F, Li L, Mazurier F, Dick JE. Generation of hematopoietic repopulating cells from human embryonic stem cells independent of ectopic HOXB4 expression. *J Exp Med.* 2005; 201:1603-1614.

[100] Narayan AD, Chase JL, Lewis RL, Tian X, Kaufman DS, Thomson JA. Human embryonic stem cellderived hematopoietic cells are capable of engrafting primary as well as secondary fetal sheep recipients. *Blood* 2006; 107: 2180-2183.

[101] Lu M, Kardel MD, O'Connor MD, Eaves CJ. Enhanced generation of hematopoietic cells from human hepatocarcinoma cell-stimulated human embryonic and induced pluripotent stem cells. *Exp Hematol* 2009; 37: 924-936.

[102] Risueño RM, Sachlos E, Lee JH, Lee JB, Hong SH, Szabo E. Inability of human induced pluripotent stem cell-hematopoietic derivatives to downregulate microRNAs in vivo reveals a block in xenograft hematopoietic regeneration. *Stem Cells* 2012; 30: 131-139.

[103] Tian X, Woll PS, Morris JK, Linehan JL, Kaufman DS. Hematopoietic engraftment of human embryonic stem cell-derived cells is regulated by recipient innate immunity. *Stem Cells* 2006; 24 (5):1370-1380.

[104] Timmermans F, Velghe I, Vanwalleghem L, De Smedt M, Van Coppernolle S, Taghon T, Moore HD, Leclercq G, Langerak AW, Kerre T, Plum J, Vandekerckhove B. Generation of T cells from human embryonic stem cell-derived hematopoietic zones. *J. Immunol.* 2009; 182 (11): 6879-6888.

[105] Woll PS, Martin CH, Miller JS, Kaufman DS. Human embryonic stem cellderived NK cells acquire functional receptors and cytolytic activity. *J. Immunol.,* 2005; 175 (8): 5095-5103.

[106] Woll PS, Grzywacz B, Tian X, Marcus RK, Knorr DA, Verneris MR, Kaufman DS, 2009. Human embryonic stem cells differentiate into a homogeneous population of natural killer cells with potent in vivo antitumor activity. *Blood* 2009; 113 (24): 6094-6101.

[107] Palis J, McGrath KE, Kingsley PD. Initiation of hematopoiesis and vasculogenesis in murine yolk sac explants. *Blood,* 1995; 86:156-163.

[108] Sabin FR. Studies on the origin of blood vessels and of red corpuscles as seen in the living blastoderm of the chick during the second day of incubation. Contrib. *Embryol.* 1920; 9:213-262.

[109] Murray PDF. The development in vitro of the blood of the early chick embryo. *Proc. Royal Soc. London*. 1932; 11: 497-521.

[110] Orkin S. GATA-binding transcription factors in hematopoietic cells. *Blood* 1992; 80: 575-581.

[111] Takakura N, Huang XL, Naruse T, Hamaguchi I, Dumont DJ, Yancopoulos GD, Suda T. Critical role of the TIE2 endothelial cell receptor in the development of definitive hematopoiesis. *Immunity* 1998; 9:677-686.

[112] Shivdasani R, Mayer E, Orkin SH. Absence of blood formation in mice lacking the T-cell leukemia oncoprotein tal-1/SCL. *Nature* 1995; 373:432-434.

[113] Shalaby F, Rossant J, Yamaguchi TP, Gertsenstein M, Wu XF, Breitman ML, Schuh AC. Failure of blood-island formation and vasculogenesis in Flk-1 deficient mice. *Nature* 1995; 376:62-66.

[114] Huber TL, Kouskoff V, Fehling HJ, Palis J, Keller G. Haemangioblast commitment is initiated in the primitive streak of the mouse embryo. *Nature* 2004; 432:625–630.

[115] Choi K, Kennedy M, Kazarov A, Papadimitriou JC, Keller G. A common precursor for hematopoietic and endothelial cells. *Development* 1998; 125:725-732.

[116] Nishikawa SI, Nishikawa S, Hirashima M, Matsuyoshi N, Kodama H. Progressive lineage analysis by cell sorting and culture identifies Flk+VE-cadherin+cells at a diverging point of endothelial hematopoietic lineages. *Development* 1998; 125: 1747-1757.

[117] Ema M, Faloon P, Zhang WJ, Hirashima M, Reid T, Stanford WL, Orkin S, Choi K, Rossant J. Combinatorial effects of Flk1 and Tal1 on vascular and hematopoietic development in the mouse. *Genes Dev.* 2003; 17:380-393.

[118] Ema M, Rossant J. Cell fate decisions in early blood vessel formation. *Trends Cardiovasc. Med.* 2003; 13:254-259.

[119] North T, Gu TL, Stacy T, Wang Q, Howard L, Binder M, Marin-Padilla M, Speck NA. Cbfa2 is required for the formation of intra-aortic hematopoietic clusters. *Development* 1999; 126:2563–2575.

[120] North TE, de Bruijn MF, Stacy T, Talebian L, Lind E, Robin C, Binder M, Dzierzak E, Speck NA. Runx1 expression marks long-term repopulating hematopoietic stem cells in the midgestation mouse embryo. *Immunity* 2002; 16:661–672.

[121] Bertrand JY, Giroux S, Golub R, Klaine M, Jalil A, Boucontet L, Godin I, Cumano A. Characterization of purified intraembryonic hematopoietic stem cells as a tool to define their site of origin. *Proc. Natl. Acad. Sci. USA* 2005; 102:134–139.

[122] Medvinsky A, Rybtsov S, Taoudi S. Embryonic origin of the adult hematopoietic system: advances and questions. *Development* 2011; 138 (6):1017-1031.

[123] Zovein AC, Hofmann JJ, Lynch M, French WJ, Turlo KA, Yang Y, Becker MS, Zanetta L, Dejana E, Gasson JC, Tallquist MD, Iruela-Arispe ML. Fate tracing reveals the endothelial origin of hematopoietic stem cells. *Cell Stem Cell* 2008; 3 (6):625-636.

[124] Kissa K, Herbomel P. Blood stem cells emerge from aortic endothelium by a novel type of cell transition. *Nature* 2010; 464 (7285):112-115.

[125] Boisset JC, van Cappellen W, Andrieu-Soler C, Galjart N, Dzierzak E, Robin C. In vivo imaging of haematopoietic cells emerging from the mouse aortic endothelium. *Nature* 2010; 464 (7285):116-120.

[126] Choi KD, Vodyanik MA, Togaratti PP,Suknuntha K, Kumar A, Samarjeet F, Probasco MD, Tian S, Stewart R, Thomson JA, Slukvin II. Identification of the hemogenic endothelial progenitor and its direct precursor in human pluripotent stem cell differentiation cultures. *Cell Rep.* 2012; 2 (3): 553-567.

[127] Ciau-Uitz A, Walmsley M, Patient P. Distinct origins of adult and embryonic blood in Xenopus. *Cell* 2000; 102 (6): 787-796.

[128] Dzierzak E, Speck NA. Of lineage and legacy – the development of mammalian hematopoietic stem cells. *Nat. Immunol.* 2008; 9: 129-136.

[129] Turpen JB, Kelley CM, Mead PE, Zon LI, 1997. Bipotential primitive-definitive hematopoietic progenitors in the vertebrate embryo. *Immunity* 1997; 7 (3): 325-334.

[130] Kennedy M, Firpo M, Choi K, Wall C, Robertson S, Kabrun N, Keller G. A common precursor for primitive erythropoiesis and definitive hematopoiesis. *Nature* 1997; 386 (6624): 488-493.

[131] Kyba M, Perlingeiro RC, Daley GQ. HoxB4 confers definitive lymphoid-myeloid engraftment potential on embryonic stem cell and yolk sac hematopoietic progenitors. *Cell* 2002; 109 (1): 29-37.

[132] Xiong JW. Molecular and developmental biology of the hemangioblast. *Dev. Dyn.* 2008; 237 (5): 1218-1231.

[133] Berens C, Hillen W. Gene regulation by tetracyclines. Constraints ofresistance regulation in bacteria shape TetR for application in eukaryotes. *Eur. J. Biochem.* 2003; 270: 3109–3121.

[134] Moscou MJ, Bogdanove AJ. A simple cipher governs DNA recognition by TAL effectors. *Science* 2009; 326 (5959):1501.

[135] Boch J, Scholze H, Schornack S, Landgraf A, Hahn S, Kay S, Lahaye T, Nickstadt A, Bonas U. Breaking the Code of DNA Binding Specificity of TAL-Type III Effectors. *Science* 2009; 326 (5959):1509-1512.

[136] Mali P, Yang L, Esvelt KM, Aach J, Guell M, DiCarlo JE, Norville JE, Church GM. RNA-guided human genome engineering via Cas9. *Science* 2013; 339 (6121):823-826.

[137] Bassett AR, Tibbit C, Ponting CP, Liu JL. Highly Efficient Targeted Mutagenesis of Drosophila with the CRISPR/Cas9 System. *Cell Rep.* 2013;4(1):220-228

[138] Shu J, Wu C, Wu Y, Li Z, Shao S, Zhao W, Tang X, Yang H, Shen L, Zuo X, Yang W, Shi Y, Chi X, Zhang H, Gao G, Shu Y, Yuan K, He W, Tang C, Zhao Y, Deng H. Induction of pluripotency in mouse somatic cells with lineage specifiers. *Cell* 2013; 153 (5): 963-975.

[139] Loh KM, Lim B. A precarious balance: pluripotency factors as lineage specifiers. *Cell Stem Cell* 2011; 8 (4):363-369.

[140] Chou BK, Cheng L. And then there were none: no need for pluripotency factors to induce reprogramming. *Cell Stem Cell* 2013; 13 (3):261-262.

Section 3

Technical Advances in the Culture and Use of Induced Pluripotent and Embryonic Stem Cells

Chapter 5

Generation and Maintenance of iPSCs From CD34+Cord Blood Cells on Artificial Cell Attachment Substrate

Naoki Nishishita, Takako Yamamoto,
Chiemi Takenaka, Marie Muramatsu and
Shin Kawamata

Additional information is available at the end of the chapter

http://dx.doi.org/10.5772/58591

1. Introduction

Cord blood (CB) cells are commonly used for the treatment of leukemia and inherited metabolic diseases. To date, more than 20,000 bone marrow transplants have been performed on children and adults with cord blood cells, and There are more than 450,000 HLA-defined CB collections stored frozen cryoperserved form in more than 50 units public CB banks and more than 2,000 CB transplants are being performed world-wide per year. CB cells are the youngest somatic cells and in theory have no post natal DNA damage such as caused by UV or chemical irritant exposure. Therefore, our previous study thought that use to the ability to cryopreserve CB HSC long-term in bank, which conferring a unique advantage to CB cell as a suitable material for generating induced pluripotent stem (iPSC) cells for future clinical use.[1]

iPSC should be generated with methods that do not require integration of exogenous DNA, thereby reducing the chance of tumorigenicity caused by random chromosomal insertion of exogenous genes. Several non-integrating reprogramming methods using EBNA based-plasmids vector [2, 3, 4, 5], piggy-back transposons [6, 7], human artificial chromosome vectors [8], small peptides [9, 10], mRNA [11] and proteins [12] have been reported. Among the vectors employed for these experiments, the Sendai virus (SeV) vector (that lacks a DNA phase) is recognized as a potent reagent for reprogramming of somatic cells [13-15]. However, complete elimination of the SeV construct carrying reprogramming factors is a key issue to assure three germ layer differentiation of individual cells. The presence of residual reprogramming factors in transfected cells could impede differentiation and contribute to formation of tumors after implantation. Therefore, the possible presence of the SeV construct should be checked at a single cell level (not at a cell clump level) utilizing single cell cloning techniques in the naïve state [16-18]. Recently, feeder-free culture systems utilizing Laminin 511, LM-E8s or Matrigel have been reported for the maintenance of established iPSCs or ES cells [19-23]. The generation of iPSCs from fibroblasts on vitronectin-coated dishes and maintenance of iPSCs in chemically defined medium on vitronectin-coated dishes has been reported [23]. These studies were to characterize as substrates that support hESCs in a sustainable undifferentiated state under a xeno-free and chemical defined culture condition [20, 23]. On the other hands, multiple matrix proteins, such as laminin, vitronectin fibronectin and synthetic polymer surfaces support hESC/iPSC growth and maintenance. Most of these materials are too expensive for large-scale usage. Because, recombinants vitronectin is relatively easy to over-express and purify, we tested vitronectin in two feeder-free ES/iPS mediums. (mTeSR-1 and ReproFF2).

In this chapter, we describe the generation of iPSC clones from cord blood cells (CBCs) in feeder-free thought naïve conditions using temperature sensitive SeV vector. Additional, human naïve iPSC culturing methods using feeder-free systems and we introduce to low-cost and stable and easy maintenance culturing methods of hESC/iPSC.

2. Experimental procedures, materials and methods

2.1. Cord blood

$CD34^{+}$CBCscan be procured from Riken Bio Resource center (Riken BRC, Ibaraki, Japan) or other commercial suppliers. Alternatively $CD34^{+}$CBCscan be obtained from fresh cord blood using a mononuclear cells isolation kit (Lymphoprep TM, Cosmo Bio Co., Japan), and a human CD34 Micro Bead kit (Miltenyi Biotec, 130-046-702) or Auto Macs columns (Miltenyi Biotec, Germany) in accordance with the manufacturer's instruction. $CD34^{+}$CBCs were cultured in the density of 1.0×10^{5} cells in two mL of hematopoietic culture medium [serum-free X-Vivo10 containing 50 ng/mL IL-6 (Peprotech, London, UK), 50 ng/mL sIL-6R (Peprotech) 50 ng/mL SCF (Peprotech), 10 ng/mL TPO (Peprotech), and 20 ng /mL Flt3/4 ligand (R&D System, Bostone, USA)] for one day prior to viral infection [23].

2.2. Preparation of coated dish for feeder-free generating iPS cells

PronectinF$^{plus®}$ coated-dish for reprogramming of CBCs is prepared as follows: One mg/mL stock solution PronectinF$^{plus®}$ (hereafter, Pronectin F, Sanyo Chemical Industries, Japan) was prepared by adding one mL of 37 °C deionized water to lyophilized Pronectin F. Ten ug/mL of Pronectin F working solution was prepared by diluting the stock solution with phosphate buffered saline (PBS). The culture dish (BD Life Science, Canada) was covered completely with Pronectin F and left overnight at room temperature. The coating solution was then removed by aspiration., and then dish was rinsed twice with PBS.

To make vitonectin-coated culture dish, the vitronectin-N (VTN-N) (Life Technology,USA) is used for a six-well plate. Dilute thawed VTN-N with 1xPBS (Life Technology,USA). in accordance with the manufacturer's instruction. Keep coated wells in culture medium at 37°C, 5% CO_2 during passaging procedure until cells are ready to be re-plated.

2.3. Sendai virus infection and reprogramming

Temperature-sensitive Sendai virus vector constructs inserting four reprogramming factors (SeV18+HS-*OCT3/4*/TS⊿F, SeV18+HS-*SOX2*/TS⊿F, SeV18+HS-*KLF4*/TS⊿F, SeV(*HNL)c-MYC*/TS15⊿F, SeV18+*GFP*/TS⊿F) were supplied by DNAVEC Corp. 1.0 x 10^4 $CD34^+$CBCs were transferred to one well of a 96-well plate in 180 μL of hematopoietic cell culture medium with 20 μL of viral supernatant containing 20 M.O.I. each of SeV constructs at 5% CO_2, 37 °C. The medium was changed to fresh medium in the following days (15-18 hours after infection). Infected cells were cultured another three days in hematopoietic culture medium in 96-well plates, after which 1 x 10^4 infected CBC were seeded on a Pronectin F-coated 6-well dish in primate ES cell medium ReproFF2 supplemented with 5 ng/mL bFGF (ReproCELL Inc, RCHEMD006B, JAPAN) to generate ES cell-like colonies under 20% O_2, 37 °C conditions. The amount of SeV constructs in the transfected cells was reduced by incubation cells at 5% CO_2, 38 °C for three days.

2.4. Cell culture in naïve state

After heat treatment, three hundred cells were resuspened in 100ml of naïve cell culture medium (see below). The cells were seeded in ten well of 96-well plate (100μl/well) pre-coated with Pronectin F. Approximately single cell in every three wells was seed in a 96-well plate. The presence of a single cell per individual well was verified by microscopic observation (phase contrast Olympus CKX31) in the same manner as single cell cloning. These cells were cultured at 37 °C in 5% O_2, 5% CO_2 condition in naïve cell culture medium to form dome-shape colonies. 50 mL of naïve ES/iPS cell culture medium was prepared by mixing 24 mL DMEM/F-12 medium (Invitrogen, 11320, Osaka), 24 mL Neurobasal medium (Invitrogen, 21103), 0.5 mL x100 nonessential amino acids (Invitrogen, 11140), 1 mL B27 supplement (Invitrogen; 17504044), and 0.5 mL N_2 supplement (Invitrogen; 17502048). The medium also contained final concentrations of 0.5 mg/mL BSA Fraction V (Sigma, A8412, Nebraska), penicillin-streptomycin (final x 1, Nacalai, Kyoto), 1 mM glutamine (Nacalai), 0.1 mM β-mercaptoethanol (Invitrogen 21985), 1 μM PD0325901 (Stemgent, 04-0006, Cambridge), 3 μM CHIR99021 (Stemgent, 04-0004), 10 μM Forskolin (Sigma, F6886) and 20 ng/mL of recombinant human LIF (Millipore; LIF1005, Billerica).

2.5. Gene chip analysis

Total RNAs from several established iPSCs lines (prime [1st, 2nd] and naïve), khES-1 (Riken BRC) and $CD34^+$CBCs (Riken BRC) were purified with an RNeasy Mini kit (QIAGEN), amplified Ovation Pico WTA System (Takara cat#3300–12), labeled with an Encore Biotin Module (Takara catalog number 4200–12) and then hybridized with a human Gene Chip (U133 plus 2.0 Array Affymetrix).

2.6. Karyotype analysis

After the iPS cells have reached the 80% of confluence, it must be harvested and fixed to make a cytogenetic suspension. iPS cells are growth arrested and accumulated in metaphase or prometaphase by inhibiting tubulin polymerization and thus preventing the formation of the mitotic spindle using colcemid (Sigma, #D7385). Following exposure to colcemid, iPS cells are treated with a hypotonic solution to enhance the dispersion of chromosomes and fixed with carnoy fixative (Methanol: Acetic Acid=3:1). Once fixed, the cytogenetic preparation can be stored in cell pellets, under fixative conditions and 20ºC for several months. Fixed cells are spread on slides and air-dried, to be finally banded for the correct identification of chromosomes.

3. Results

3.1. Selection of coating materials for feeder-free generating iPS cells

Using gene chip approach, we investigated the levels of adhesion molecule expression on (i) CD34⁺CBCs, (ii) the resulting iPSC cells and (iii) naïve iPSC on SNL (SNL76/7, ECACC) cultured in naïve cell medium. We identified several molecules that were expressed by CD34⁺CBCs and by the resulting primed and naïve iPSCs cultured on feeder cell SNL (Table 1). These data prompted us to use their ligands to anchor CBCs to dishes for reprogramming in a feeder-free system. In this context, fibronectin or a relevant material, which has an-Arg-Gly-Asp-(RGD) motif that can bind to the integrin $\alpha5/\beta1$ dimmer expressed on CBCs, was selected as a candidate for a coating material for the generation of iPSCs.

Gene ID	Description	Cord blood cell (CBC)	iPS cell from CD34⁺ on SNL cells	Naïve state iPSCs on SNL cells
		CD34⁺ (n=3)	SeV-CB iPS (n=3)	Naïve state SeV-CB-iPS (n=3)
COL1A1	Collagen, type I, alpha 1	6.87 ± 3.89	21,022.23 ± 1,3691.11	18,154.75 ± 6,733.33
COL1A2	Collagen, type I, alpha 2	9.15 ± 7.69	4,170.04 ± 1,796.98	22,075.02 ± 9,436.53
COL9A3	Collagen, type IX, alpha 3	5.20 ± 2.01	1,793.79 ± 452.86	756.78 ± 130.64
COL18A1	Collagen, type XVIII alpha 1	132.90 ± 73.98	2,565.17 ± 877.84	1,558.63 ± 897.07
ITGA5	Integrin alpha 5	284.25 ± 38.21	366.91 ± 24.61	372.02 ± 22.96
ITGB1	Integrin, beta 1 (ITGB1)	9,034.55 ± 2,178.51	22,946.52 ± 7,287.56	25,447.85 ± 7,076.74
SDC2	Syndecan 2	132.13 ± 20.79	6,224.00 ± 813.19	1,095.14 ± 651.86
SDC4	Syndecan 4	334.2 ± 161.06	1,973.21 ± 565.5	11,594.85 ± 6,677.37
FN1	Fibronectin1	7.92 ± 1.14	9,922.15 ± 3,769.53	14,929.00 ± 6,824.63
TJP1	Tight junction protein 1	53.17 ± 23.59	8,351.35 ± 1,682.14	6,850.59 ± 801.16
TJP2	Tight junction protein 2	1,056.53 ±309.49	3,023.60 ± 59.05	4,144.96 ± 1,675.39

Mean and standard deviation of signal values for the expression of indicated genes from three independent experiments.

Table 1. Gene chip analysis of adhesion molecules on CD34⁺cells, and primed and naive iPSCs cultured on SNL.

From the point of view of quality control and reagent tracking, synthetic peptides expressing the RGD motif are preferable to natural ligands. Thus, Pronectin F which mimics the peptide

http://dx.doi.org/10.5772/58591

structure of fibronectin, was chosen and tested for reprogramming CBCs. Pronectin F was synthesized by fusing two amino acid motifs, RGD and (GAGAGS)^9 in tandem to produce a-RGD-(GAGAGS)^9-RGD-(GAGAGS)^9-RGD-(GAGAGS)^9-RGD-polypeptide. This polypeptide has thirteen RGD motifs and is folded at the RGD sequence. Thus, the RGD motif is effectively exposed at the limbs of the peptide bundle, facilitating its potent binding affinity to the integrin α5/β1 dimer.

3.2. Generation of iPS cells on synthetic peptide (Pronectin F®)

Protocol for generating iPSC on feeder less condition is shown in Figure 1.

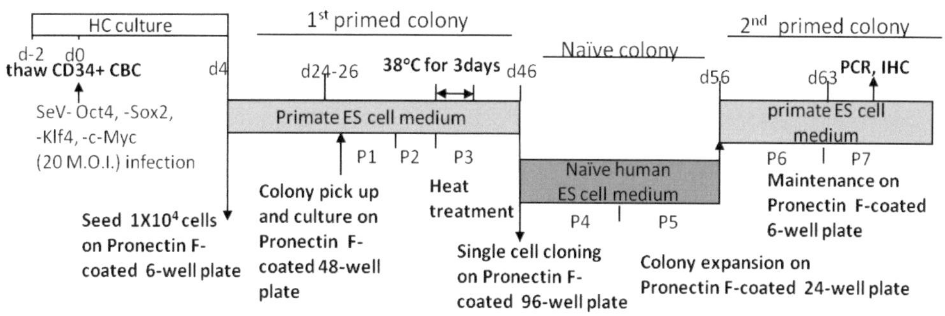

Figure 1. Protocol for generation of iPSCs from CD34$^+$CBCs on Pronectin F-coated dishes with temperature sensitive SeV vectors. P: passage.

Human ES cell-like colonies (first prime state) were picked up at day 24 and cultured on Pronectin F-coated dishes. The colonies were subjected to heat treatment (38°C, three days) at passage three (P3). Colonies emerged from single cells in Pronectin F-coated 96-well plates under naïve conditions at P4, dome-shaped colonies at P5 under naïve conditions, ES cell-like colonies (second primed) cultured under primed culture conditions at P6,P7.

The medium was changed every other day for transformed adherent cell stage (day 1-12). However, during day 13-17, primate ES medium was changed every day. The reprogramming process was monitored by checking the morphology of the transfected cells. CD34+cells infected with SeV constructs were cultured in serum-free hematopoietic cell culture, as shown in Figure 2 (day1). Some cells attached to Pronectin F-coated dishes by day four in Figure 2 (day 4). Cobble stone-like cell colonies emerged at day nine and cell clumps with round and small cells emerged inside the colonies at day 13 on Pronectin F-coated dishes (Figure 2, day 9, day 13). Cell clumps within cobble stone-like colonies grew (Figure 2, day 17) and finally human ES cell-like colonies emerged (Figure 2, day 24) on Pronectin F-coated dishes which were then picked up for serial passage. Fifteen to twenty-two dish-shape human ES cell-like colonies were picked out of 10,000 CD34+CBCs seeded on Pronectin F-coated dish in primate ES medium. Colonies were picked approximately three weeks after viral infection. Cells from individual colonies were transferred to a Pronectin F-coated 48-well plate to select passage-able ES cell-like colonies capable of passage.

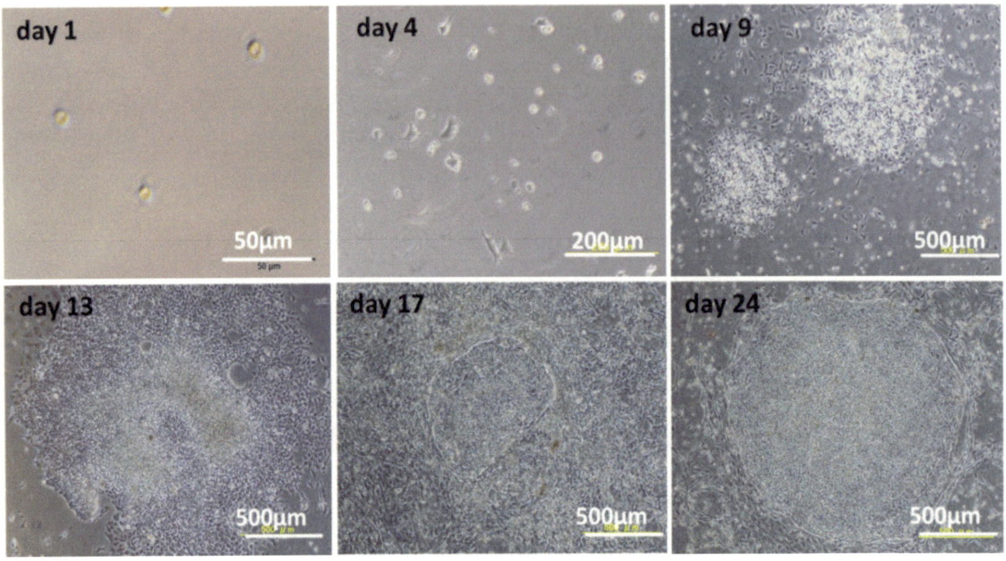

Figure 2. Phase contrast light microscopic observation of cells during reprogramming. Images captured on a Pronectin F-coated dish on days 1, 4, 9, 13, 17 and 24. day 1; Infected CD34⁺CBCs with Sendai virus seed on Pronectin F⁺-coated dish. day 4; Infected CD34⁺CBCs were attachment and little spread on Pronectin F-coated dish. day 9: Infected CD34⁺CBCs expansion on Pronectin F-coated dish. day 13; Infected CD34⁺CBCs expansion with colony-like state. day 17; generation of small ES-like colony around spreading apart of infected CD34⁺CBCs. day 24; Human ES cell-like colonies emerged on Pronectin F-coated dishes.

Human ES cell-like colonies (first primed state) were picked up at day 24 and cultured on Pronectin F-coated dishes. The colonies were subjected to heat treatment (38 °C, three days) at passage three (P3) to reduce the SeV constructs (Figure 3).

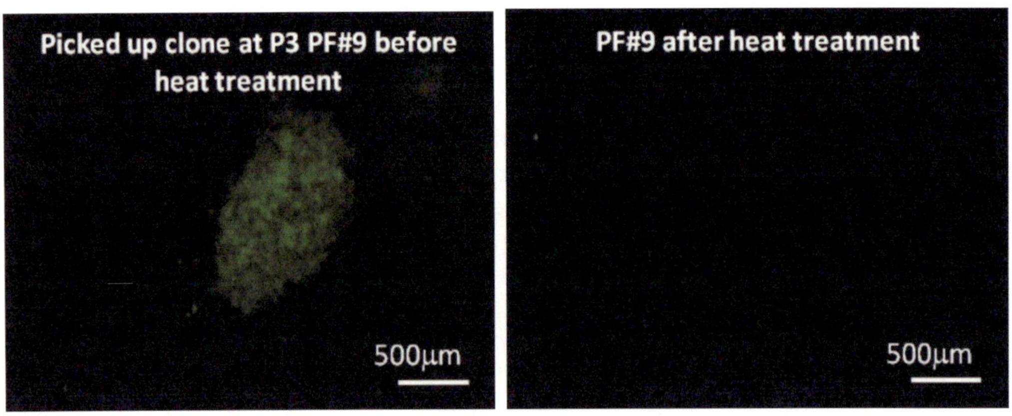

Figure 3. Expression of SeV in reprogrammed cell clone before and after heat treatment.

Expression of SeV in ES cell-like colonies before heat treatment at passage three (SeV at P3) and after heat treatment and single cell cloning at passage.

Reprogrammed cell clone before single cell cloning in the naïve state was named **PF** (**P**ronectin F –coated **F**eeder-less clones). The level of SeV protein expression was determined by immunostaining with SeV HN antibody (polyclonal-rabbit, gift to DNAVEC Corp., Ibaraki).

Then, single cells from dish-shaped (first primed) primate ES cell-like colonies at passage three were seeded on a Pronectin F-coated 96 well plate at approximately one cell per three wells and cultured in naïve medium under hypoxic conditions (5% O_2, 5% CO_2 at 37° C). After five or six days, dome-shaped mouse ES cell like-colonies were collected and expanded on Pronectin F-coated dishes. Next, cell clumps were transferred to primate ES medium under 20% O_2 again to culture them in the primed state in Figure 4.

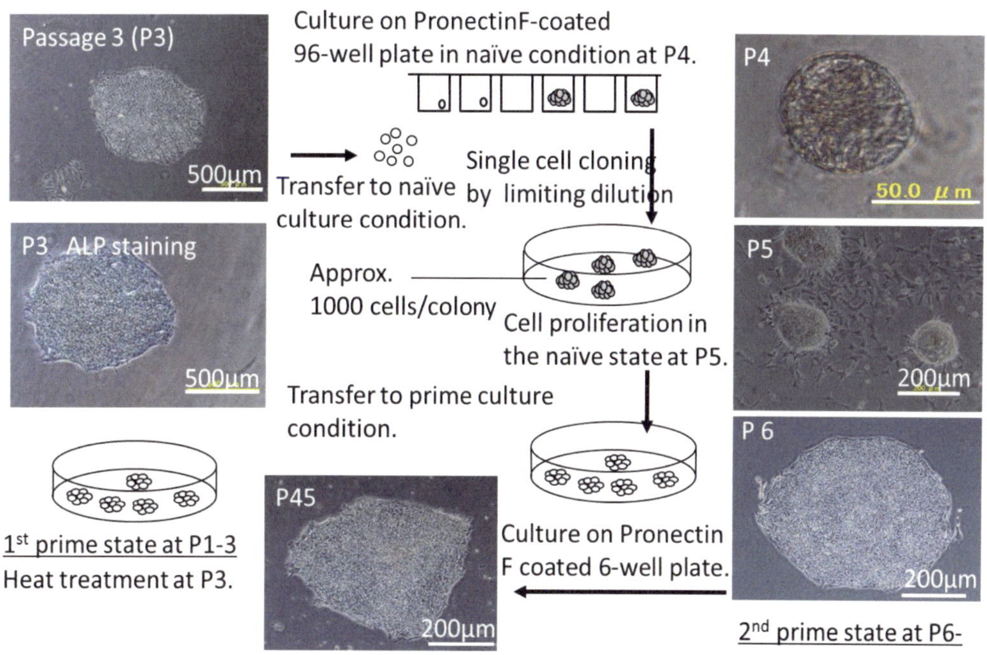

Figure 4. Generation of reprogrammed cell clone from a single cell via the naïve state.

Light microscopic image and ALP staining at P3 are shown in upper and lower panels, respectively. Colonies emerged from single cells in Pronectin F-coated 96-well plates under naïve condition at P4, dome-shaped colonies at P5 under naïve condition, ES cell-like colonies (second primed) cultured under primed culture condition at P6 or long-term passaged clone at P45 are shown. And, the colonies were subjected to heat treatment (38°C, three days). Colonies emerged from single cells in Pronectin F-coated 96-well plates under naïve conditions at P4, dome-shaped colonies at P5 under naïve conditions, ES cell-like colonies (second primed).

Long-term passaged clone (PFX#9) at P45 is shown. After single cell cloning in the naïve state, picked up cell clones were named as **PFX** (**P**ronection **F**-coated **F**eeder-free iPSC derived from female (**XX**) cord blood cell. We used female cord blood cells (XX) to check the status of being in the naïve stage manifested by reactivation of X-chomosome inhibition. Culturing cells in the naïve state was useful for a single cell cloning in limited dilution, but we fail to support cell culture in the naïve stage robustly for more than five passages. Therefore cells were kept culturing in the primed condition (20% O_2, the ES cell medium containing bFGF) after single cell cloning in the naïve state for further appraisal and passages.

Whether dome–shape cells cultured in the naïve condition (Figure 4, P5) was indeed in the naïve state or not a reactivation of X-chromosome inhibition was determined by gene chip analysis (Table 2.) and RT-PCR (Figure 5) with each states (prime [1st, 2nd] and naïve).

probe no.	Gene description	khES-1*(XY)* 1st *primed*	13PFX #1*(XX)* 1st primed	PFX#2 *(XX)* naïve	PFX#7 *(XX)* naïve	PFX#9 *(XX)* naïve	PFX#2 *(XX)* 2nd primed	PFX#7 *(XX)* 2nd primed	PFX#9 *(XX)* 2nd primed
214218_s_at	*Xist*	23.19	2870.9	40.24	39.2	47.7	1,610.52	2,615.94	1,278.82
221728_x_at	*Xist*	46.12	6727.3	5.53	71.2	75.7	2,949.19	5,036.44	2,275.75
224588_at	*Xist*	20.47	10579.6	8.09	17.4	48.2	2,891.72	6,009.63	2,604.95
227671_at	*Xist*	65.22	5964.9	3.24	76.5	79.3	2,761.26	5,532.28	2,217.70

Table 2. *Xist* gene expression analysis by gene chip for X-chromosome activite / inactivite states using four different probes.

Naïve PFXs were cultured in the naïve state and 2nd primed PFs were cultured in the naïve state. PF #13 1st prime and khES-1 1st primed were cultured in the primed state (without being in the naïve state). PF #13 and PFXs are female (XX) in origin, while human ES cell line khES01 is male in (XY) origin.

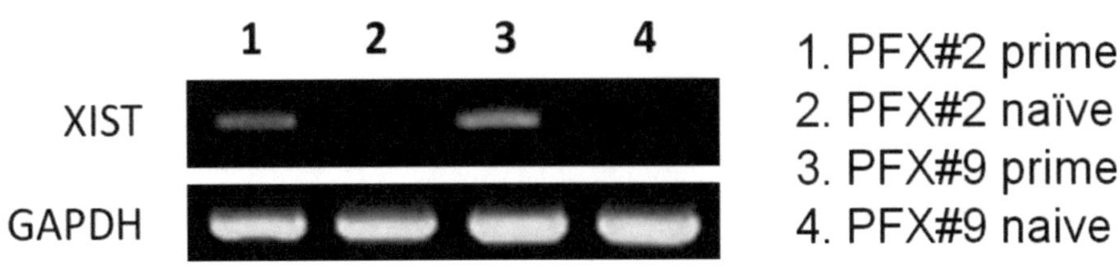

Figure 5. Expression of *Xist* genes in naïve and prime state iPS cell determined by RT-PCR.

RT-PCR determination of naïve state in iPSC colony, second primed colonies 1; PFX#2 or 3; PFX#9, naïve state colonies before each second prime state colonies (2; PFX#2 naïve, 4; PFX#9 naïve). Values were normalized using the housekeeping gene *GAPDH*.

3.3. Maintenance and characterization of reprogrammed cells

3.3.1. Maintenance of reprogrammed cells established in feeder free condition

Once established on a Pronectin-coated dish, reprogrammed cell colonies can be maintained either in a Pronectin F-, Laminin-or Matrigels-coated dish for serial passage. 100-200 cell clumps (50-100 μm diameters) were seeded on 100mm dish or in six wells of a 6-well plate and cultured until colonies reach 70-80% confluence. The split ratio was routinely 1:3. This is a protocol for passage via cell clump, not via single cell suspension.

It is possible to conduct cell passaging via single cell suspension in serum-free media (mTeSR1, TeSR2 and ReproFF) in the primed condition with the use of Rock inhibitor (Y-27632, Stemgent, #2514).

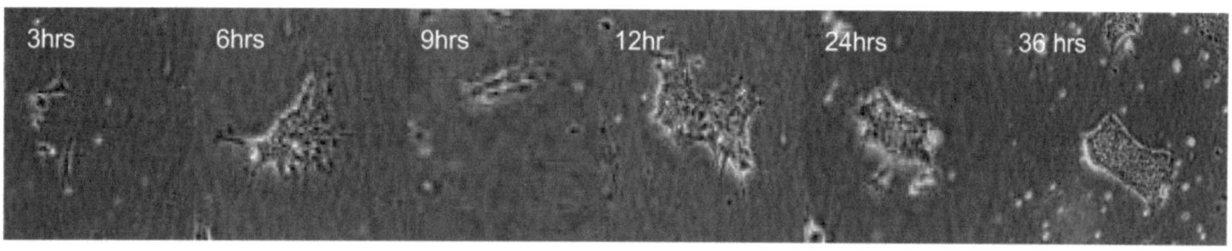

Figure 6. Photograph of forming cell colony from single iPS cell suspension on Matrigel.

As shown in Figure 6, it is notable that single cells migrate towards one another three to thirty-six hrs after passage to form cell clumps. This is a single cell passage, not a single cell cloning process. We failed to generate colonies from single cell in the primed state. That is a rationale for using naïve culture for single cell subcloning purpose. It is convenient to use singe cell suspension for passage. However, morphology of cell colony via single cell passage in longer period (P20 or over) is not uniform and is no longer round. We have not accumulated enough data how relevant this even is, but from a daily practical point of view, we perform cell passaging via cell clumps.

3.3.2. Characterization of reprogrammed cells by Reverse transcriptase polymerase chain reaction (RT-PCR)

The expression of pluripotecy related genes were determined by RT-PCR. Total RNA was purified with an RNeasy Plus Micro kit (QIAGEN 74034), according to the manufacturer's instructions, and One μg of total RNA was used for reverse transcription reactions with PrimeScript RT reagent kit (TAKARA, Japan). Result is shown in Figure 7. Primer sequences used for PCR are shown in Table 3.

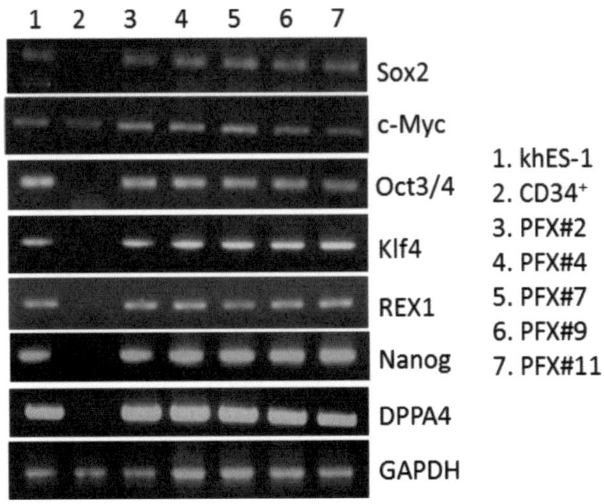

Figure 7. Expression of endogenous pluripotency related genes in reprogrammed cell determined by RT-PCR.

Primers		Size(bp)	
hOCT3/4-F1165	GAC AGG GGG AGG GGA GGA GCT AGG	Undifferentiated ES cell (endo)	144
hOCT3/4-R1283	CTT CCC TCC AAC CAG TTG CCC CAA AC		
hSOX2-F1430	GGG AAA TGG GAG GGG TGC AAA AGA GG	undifferentiated ES cell (endo)	151
hSOX2-R1555	TTG CGT GAG TGT GGA TGG GAT TGG TG		
hMYC-F253	GCG TCC TGG GAA GGG AGA TCC GGA GC	undifferentiated ES cell (endo)	328
hMYC-R555	TTG AGG GGC ATC GTC GCG GGA GGC TG		
hKLF4-F1128	ACG ATC GTG GCC CCG GAA AAG GAC C	undifferentiated ES cell (endo)	397
hKLF4-R1826	TGA TTG TAG TGC TTT CTG GCT GGG CTC C		
DPPA4-F	GGAGCCGCCTGCCCTGGAAAATTC	undifferentiated ES cell	408
DPPA4-R	TTT TTC CTG ATA TTC TAT TCC CAT		
hTERT-F3292	CCT GCT CAA GCT GAC TCG ACA CCG TG	undifferentiated ES cell	445
hTERT-R3737	GGA AAA GCT GGC CCT GGG GTG GAG C		
REX1-F	CAG ATC CTA AAC AGC TCG CAG AAT	undifferentiated ES cell	306
REX1-R	GCG TAC GCA AAT TAA AGT CCA GA		
NANOG-F	CAG CCC CGA TTC TTC CAC CAG TCC C	undifferentiated ES cell	391
NANOG-R	CGG AAG ATT CCC AGT CGG GTT CAC C		
SeV vector-F	GGATCACTAGGTGATATCGAGC	SeV vectors	193
SeV vector-R	CATATGGACAAGTCCAAGACTTC		
hGAPDH F	AAC AGC CTC AAG ATC ATC AGC	control	337
hGAPDH R	TTG GCA GGT TTT TCT AGA CGG		

Table 3. List of genes and the primers used for RT-PCR.

3.3.3. *Evaluation for remaining SeV construct*

The remaining SeV construct after heat treatment and single cell cloning was determined by qRT-PCR and shown in Table 4.

Cell	$CD34^+$ infected	PF#7 before HT	PF#9 before HT	$CD34^+$	201B7	PFX#7 (P8)	PFX#9 (P8)
SeV	193736	19719	26850	6511	3997	5414	1135

Table 4. Quantitative RT-PCR determination of residual SeV viral genomes.

Quantitative RT-PCR determination of residual SeV viral genomes in $CD34^+$CBCs three days after SeV infection (CD34 infected), first primed colony iPS#7 or iPS#9 before heat treatment at P2 (PF#7 before HT, PF#9 before HT), non-infected $CD34^+$CBCs (CD34) or iPSC clone generated by retrovirus (201B7), established clones at P9 (PFX#7) or (PFX#9). Values were normalized using the housekeeping gene *GAPDH*.

The residual SeV viral genome was determined by qRT-PCR analysis for selection of non-integration and non-virus of established iPSC lines.

3.3.4. *Characterization of reprogrammed cells by Immunohistological staining*

ES cell like-colonies were stained with the Leukocyte Alkaline Phosphatase kit (VECTOR, Burlingame, CA, SK-5300) in accordance with the manufacturer's instructions. For immunochemical staining, these cells were fixed with 4% paraformaldehyde followed by staining with antibodies against Oct3/4 (1:100 sc-5279; Santa Cruz, Biotechnology USA), Nanog (1:500, RCAB0003P; Reprocell, Tokyo, Japan), SSEA-3 (1:200 MAB4303; Millipore), SSEA-4 (1:200, MAB4304, Millipore). Photomicrographs were taken with a fluorescent microscope (Olympus BX51, IX71, Tokyo) and a light microscope (Olympus CKX31).

ES cell-like clone PFX#9 at P8 was stained with antibodies against Nanog, Oct3/4, SSEA-3, or SSEA-4 as indicated. Alexa 594-and Alexa 488-conjugated secondary antibodies (red and green, respectively) were used to visualize the staining.

3.3.5. *Characterization of reprogrammed cells by gene chip analysis and karyotyping*

Total RNAs from several established iPSCs lines, ESCs lines (Riken BRC) and CD34+CBCs (Riken BRC) were purified with an RNeasy Plus Mini kit (QIAGEN 74136), amplified Ovation Pico WTA System (Takara cat#3300-12), labeled with an Encore Biotin Module (Takara catalog number 4200-12) and then hybridized with a human Gene Chip (U133 plus 2.0 Array Affymetrix) according to the manufacturer's instructions (Figure 9). Karyotyping G-band method of iPSCs is shown in Figure 10. The amount of metaphases obtained is sometimes inadequate for chromosome analysis, thus it is always necessary to keep growing the PFX#9 iPS cells. As shown in Figure 10, PFX#9 iPS cell on VTN-N was normal karyotypic cell.

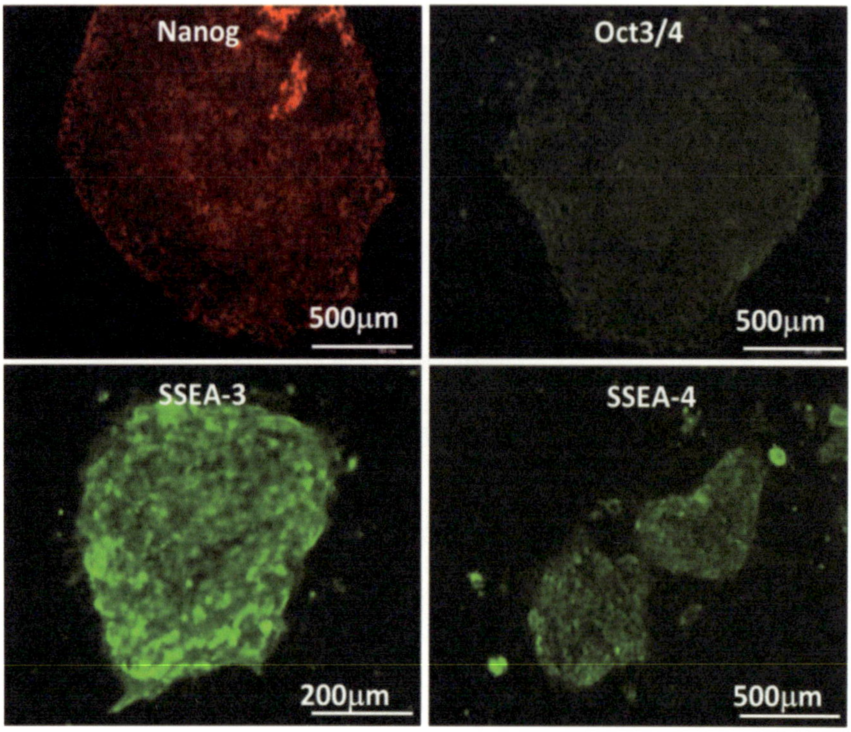

Figure 8. Characterization of established iPSC PFX#9 clones. Expression of pluripotency-related molecules in reprogrammed cell clones.

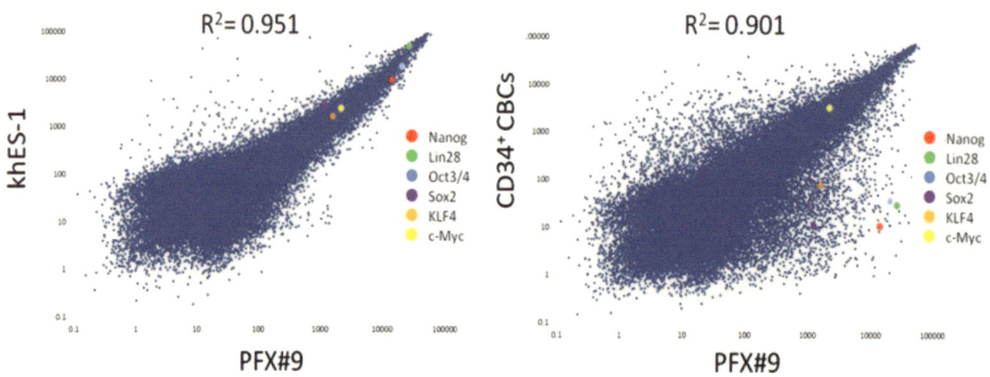

Figure 9. Gene expression comparison between the mean (mean) expression of clustered pluripotent stem cell. [PFX#9(iPSC from CBC with Yamanaka 4factors-heat treat Sendai virus without feeder) and HSC of Cord Blood ($CD34^+$CBC)] and gene expression of PFX#9, or that of khES-1, (iPSC from CBC with Yamanaka 4 factors-Sendai Virus on feeder) ((left panel)]. R^2: dicision coefficient

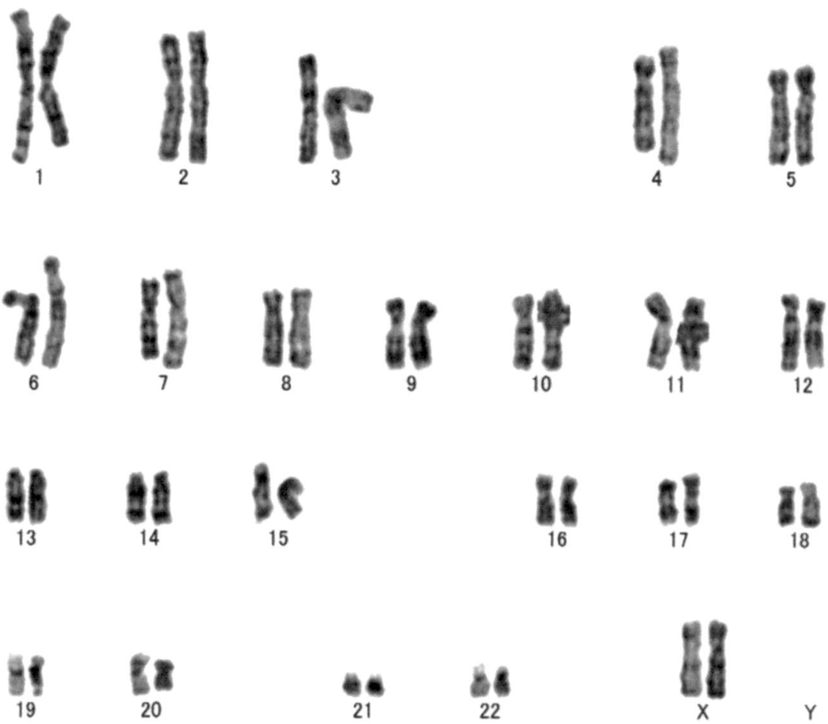

Figure 10. G-band karyotype analyses. PFX#9 (P45)

3.3.6. In vitro differentiation potentials of reprogrammed cells

The three germ layers differentiation potential of reprogrammed cells was tested via embryo body (EB) formation. Established ES cell-like clones were transferred to six-well, ultralow attachment plates (Corning) and cultured in DMEM/F12 containing 20% knockout serum replacement (KSR, Invitrogen) 2 mM L-glutamine, 1% NEAA, 0.1 mM 2-ME and 0.5% penicillin and streptomycin or ReproFF medium without bFGF to form EB. The medium was changed every other day. The resulting EBs were transferred to gelatin-coated plates for 16 days. Differentiation to ectodermal, mesodermal, or endodermal tissue was confirmed by detection of molecules related to three germ layers lineage differentiation such as α-feto-protein (endoderm), βIII-tubulin (ectoderm), GFAP (ectoderm), or Vimentin (mesoderm) with antibody against α-feto-protein (1:100 dilution MAB1368; R&D Systems), βIII-tublin (1:200 T4026; Sigma), GFAP(1:50 sc-6170 santa cruz biotechnology) or Vimentin (1:100 sc-5565; Santa Cruz Biotechnology) respectively. Antibodies were visualized with Alexa Fluor 488 goat anti-mouse (1:1,000; Invitrogen), Alexa Fluor 594 rabbit anti-mouse (1:1,000; Invitrogen), and Alexa Fluor 594 goat anti-rabbit (1:1,000; Invitrogen). Nuclei were stained with DAPI (1:1,000; Sigma) as shown in Figure 11.

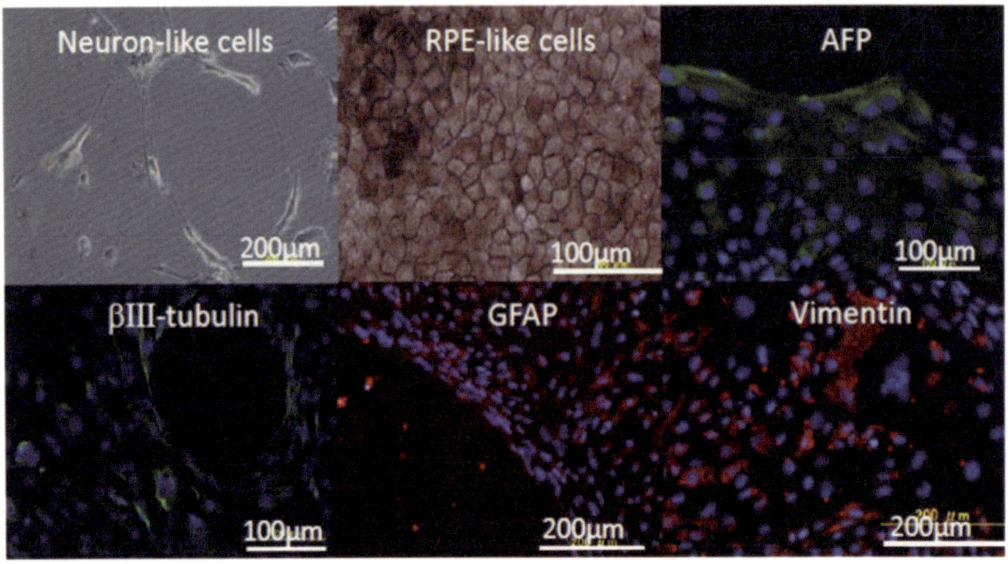

Figure 11. *In vivo* differentiation potential of established clones. Phase contrast images of neuron-like (top left) and retinal pigmented epithelium (RPE) differentiation (top middle) of established clone PFX#7. Cells were fixed and stained with antibodies against AFP, βIII-tubulin, GFAP and vimentin to identify specific cell lineages [18].

3.3.7. In vivo differentiation potential of reprogrammed cells by Teratoma formation assay

Reprogrammed cell lines should demonstrate differentiation potential reflecting three germ layers, *in vivo* as well as *in vitro*. To this end, one million iPSCs were injected beneath the testicular capsule of NOD-SCID mice (SLC Japan) to determine the ability of the transplanted cells to form teratomas containing cells of all three germ layers. Tumor formation was observed approximately four weeks after cell transplantation. Tumor tissues were fixed with 4% formalin, sectioned, and stained with hematoxylin and eosin (Figure 12).

3.3.8. Preservation of Feeder-free iPS cells

Human ES/iPS clones generated and maintained in a feeder-free system could be frozen in cell clumps using DMSO-free, chemically defined and serum-free freezing medium, CryoStem™ Freezing Medium (Stemgent), and could be cultured again on a Pronectin F-coated dish after thawing. Approximately 10-20% of the colony number scored before cryopreservation in CryoStem™ emerged after thawing.

3.3.9. Long-term, Low-cost and Stable maintenance of undifferentiated human induced pluripotent stem cells in feeder-free condition

Vitronectin provides a completely defined culture system for the maintenance of hiPSC under feeder-free conditions such as ReproFF2 medium (Figure 13, Figure 14, Table 5). This system

http://dx.doi.org/10.5772/58591

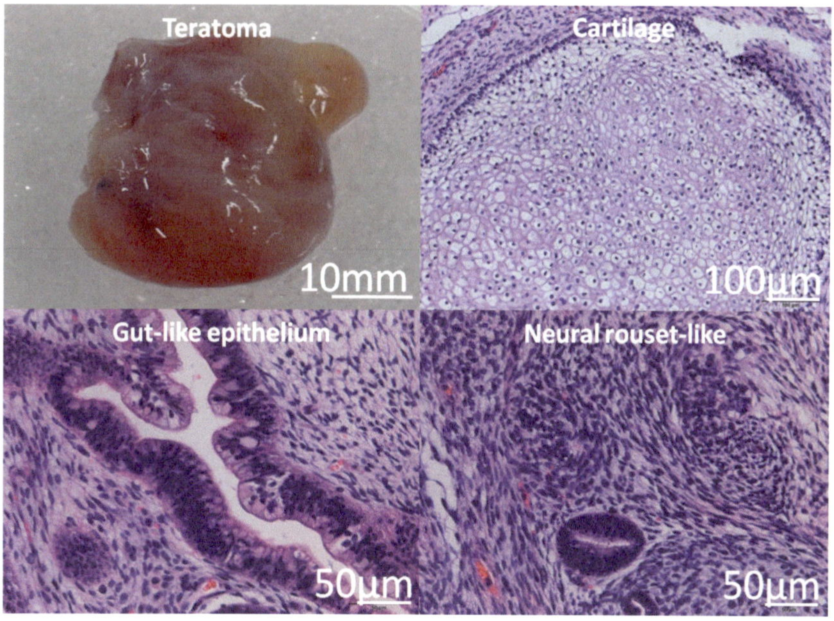

Figure 12. Teratoma with cystic structure. It was derived from iPSCs (PFX #9) implanted in the testicular capsule of a NOD-SCID mouse. It was stained with hematoxylin and eosin for histological observation.

allows complete control over the culture environment, resulting in more consistent cell populations and reproducible results in clinical applications.

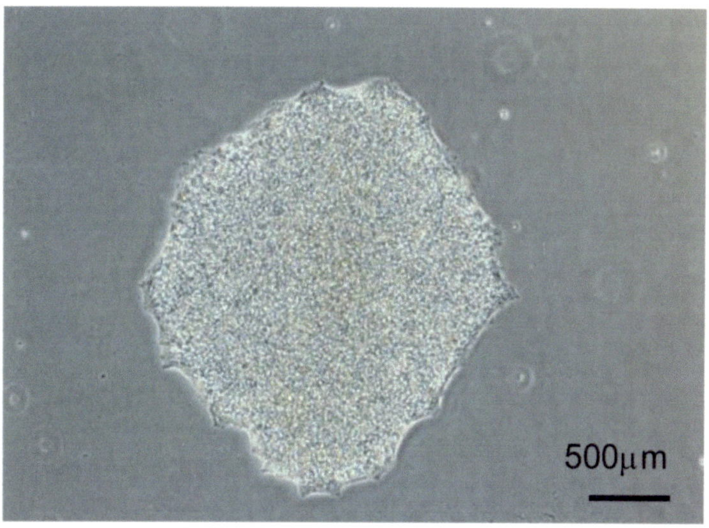

Figure 13. Maintenance of iPS cells (PFX#9) on recombinant vitronectin (VTN-N, Life Technology) in ReproFF2 medium.

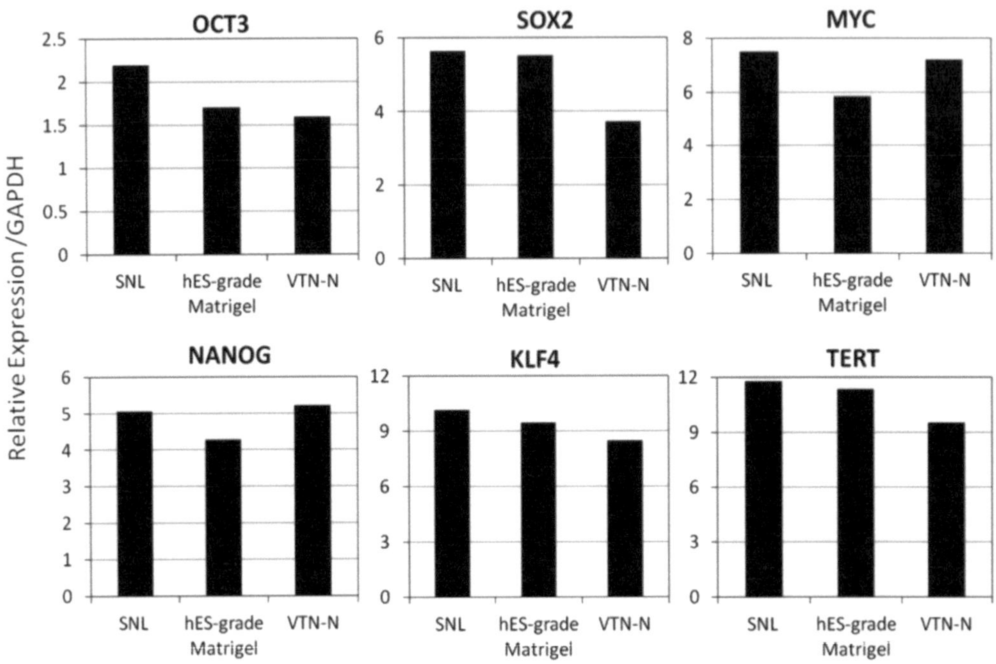

Figure 14. Expression of endogenous pluripotency related genes in iPSC (PFX#9) on VTN-N determined by qRT-PCR.

Following, the PFX#9 cells cultured with VTN-N was the gene expression of pluripotency markers comparable iPS cells cultured on Matrigel or on SNL in Figure 12. It was found that only a recombinant vitronectin (VTN-N) can be maintained in culture for long-term feeder free conditions.

Primers		Size(bp)	
hOCT3/4-F	GAA ACC CAC ACT GCA GCA GA	undifferentiated ES cell	103
hOCT3/4-R	TCG CTT GCC CTT CTG GCG		
hSOX2-F	GGG AAA TGG GAG GGG TGC AAA AGA GG	undifferentiated ES cell	151
hSOX2-R	TTG CGT GAG TGT GGA TGG GAT TGG TG		
hMYC-F	CGT CTC CAC ACA TCA GCA CAA	undifferentiated ES cell	68
hMYC-R	TCT TGG CAG CAG GAT AGT CCT T		
hKLF4-F	CGC TCC ATT ACC AAG AGC TCA T	undifferentiated ES cell	77
hKLF4-R	CGA TCG TCT TCC CCT CTT TG		
hTERT-F	CGT ACA GGT TTC ACG CAT GTG	undifferentiated ES cell	82
hTERT-R	ATG ACG CGC AGG AAA AAT GT		

http://dx.doi.org/10.5772/58591

Primers		Size(bp)	
REX1-F	TGC AGG CGG AAA TAG AAC CT	undifferentiated ES cell	64
REX1-R	TCA TAG CAC ACA TAG CCA TCA CAT		
NANOG-F	CTC AGC TAC AAA CAG GTG AAG AC	undifferentiated ES cell	153
NANOG-R	TCC CTG GTG GTA GGA AGA GTA AA		
hGAPDH-F	CCA CTC CTC CAC CTT TGA CG	control	114
hGAPDH-R	ATG AGG TCC ACC ACC CTG TT		

Table 5. List of primers used for quantitative real-time PCR (qRT-PCR)

4. Conclusion

In this chapter we have shown the method for generating iPSC from non-cultured CD34$^+$cord blood cells using feeder-free conditions. The established cell clones were characterized at a single cell level. This robust iPSC generation method will solve some of the safety concerns related to tumorigenicity ariseing from chromosomal integration of exogenous genes and/or infection hazards associated with the use of by xenogeneic biological products in the culture system. These methods will contribute to future application of iPSCs-derived cell therapy.

Acknowledgements

We thank SI Nishikawa for useful discussions, N Fusaki for samples gift, and S Sawada for gene chip analysis. We thank Riken Bio Resource Center (Tukuba, Japan) for fresh CD34$^+$CBCs. This work was partly supported by the Regulatory Science Study for Safety issue for pluripotent stem cells of JST Tokyo Japan (2010-2014) and Adaptable and Seamless Technology Transfer Program through Target-driven R&D (A-STEP) project of JST Tokyo Japan (2013).

Author details

Naoki Nishishita, Takako Yamamoto, Chiemi Takenaka, Marie Muramatsu and Shin Kawamata

Foundation for Biomedical Research and Innovation, Kobe, Japan

References

[1] Takenaka C, Nishishita N, Takada N, Jakt LM, Kawamata S. et al (2010) Effective generation of iPS cells from CD34+cord blood cells by inhibition of p53. Exp Hematol. 2010 38(2):154-62.

[2] Okita K, Matsumura Y, Sato Y, Okada A, Morizane A, (2011) A more efficient method to generate integration-free human iPS cells. *Nature Methods* 322: 409-12.

[3] Soldner F, Hockemeyer D, Beard C, Gao Q, Bell GW, et al. (2009) Parkinson's disease patient derived induced pluripotent stem cells free of viral reprogramming factors. *Cell* 136: 964–977.

[4] Okita K, Yamakawa T, Matsumura Y, Sato Y, Yamanaka S, et al. (2013) An efficient nonviral method to generate integration-free human-induced pluripotent stem cells from cord blood and peripheral blood cells. *Stem Cells*, 31(3):458-66.

[5] Su RJ, Baylink DJ, Neises A, Kiroyan JB, Zhang XB. et al. (2013) Efficient generation of integration-free ips cells from human adult peripheral blood using BCL-XL together with Yamanaka factors. *PLoSOne*, 21; 8(5):e64496.

[6] Woltjen K, Michael IP, Mohseni P, Desai R, Mileikovsky M, et al. (2009) piggy Bac transposition reprograms fibroblasts to induced pluripotent stem cells. *Nature* 458: 766–770.

[7] Kaji K, Norrby K, Paca A, Mileikovsky M, Mohseni P, et al. (2009) Virus-free induction of pluripotency and subsequent excision of reprogramming factors. *Nature* 458: 771–775.

[8] Hiratsuka M, Uno N, Ueda K, Kurosaki H, Oshimura M, et al. (2011) Integration-free iPS cells engineered using human artificial chromosome vectors. PLoSOne. 6(10), e25961.

[9] Li W, Wei W, Zhu S, Zhu J, Shi Y, et al. (2009) Generation of rat and human induced pluripotent stem cells by combining genetic reprogramming and chemical inhibitors. *Cell Stem Cell* 4: 16–9.

[10] Huangfu D, Maehr R, Guo W, Eijkelenboom A, Snitow M, et al. (2008) Induction of pluripotent stem cells by defined factors is greatly improved by small-molecule compounds. *Nat Biotechnol* 26: 795–7.

[11] Warren L, Manos PD, Ahfeldt T, Loh YH, Li H, et al. (2010) Highly efficient reprogramming to pluripotency and directed differentiation of human cells with synthetic modified mRNA. Cell Stem Cell, 7(5), 618-30.

[12] Kim D, Kim CH, Moon JI, Chung YG, Chang MY, et al. (2009) Generation of human induced pluripotent stem cells by direct delivery of reprogramming proteins. *Cell Stem Cell* 4: 472-476.

[13] Fusaki N, Ban H, Nishiyama A, Saeki K, Hasegawa M (2009) Efficient induction of transgene-free human pluripotent stem cells using a vector based on Sendai virus, an RNA virus that does not integrate into the host genome, *Proc Jpn Acad Ser B Phys Biol Sci.* 85: 348-62.

[14] Seki T, Yuasa S, Oda M, Egashira T, Yae K, et al. (2010) Generation of induced pluripotent stem cells from human terminally differentiated circulating T cells. *Cell Stem Cell* 7: 11-4.

[15] Ban H, Nishishita N, Fusaki N, Tabata T, Saeki K et al. (2011) Efficient generation of transgene-free human induced pluripotent stem cells (iPSCs) by temperature-sensitive Sendai virus vectors. *PNAS* 108: 14234-14239.

[16] Guo G, Yang J, Nichols J, Hall JS, Eyres I, et al. (2009) Klf4 reverts developmentally programmed restriction of ground state pluripotency. *Development* 136(7): 1063–1069.

[17] Hanna J, Cheng AW, Saha K, Kim J, Lengner CJ, et al. (2010) Human embryonic stem cells with biological and epigenetic characteristics similar to those of mouse ESCs. *PNAS* 107(20): 9222-7.

[18] Nishishita N, Shikamura M, Takenaka C, Takada N, Fusaki N, et al. (2012). Generation of Virus-Free Induced Pluripotent Stem Cell Clones on a Synthetic Matrix via a Single Cell Subcloning in the Naïve State. *PLoSOne* e38389.

[19] Zhou H, Li W, Zhu S, Joo JY, Do JT, et al. (2010) Conversion of mouse epiblast stem cells to an earlier pluripotency state by small molecules. *J Biol Chem* 285(39): 29676-80.

[20] Miyazaki T, Futagi S, Suemori H, Sekiguchi K, Kawase E, et al. (2012), Laminin E8 fragments support efficient adhesion and expansion of dissociated human pluripotent stem cells, *Nature communications*, 3:1236.

[21] Rodin S, Domogatskaya A, Ström S, Hansson EM, Chien KR, et al. (2010) Long-term self-renewal of human pluripotent stem cells on human recombinant laminin-511. *Nat Biotechnol*. 28(6):611-5.

[22] Hakala H, Rajala K, Ojala M, Panula S, Areva S, et al (2009) Comparison of biomaterials and extracellular matrices as a culture platform for multiple, independently derived human embryonic stem cell lines. *Tissue Eng. part A* 15(7):17775-85.

[23] Chen G, Gulbranson DR, Hou Z, Bolin JM, Ruotti V, et al (2011) Chemically defined conditions for human iPSC derivation and culture, *Nature Methods*(5) 424-429.

Chapter 6

Cadherin-Fc Chimeric Protein-Based Biomaterials: Advancing Stem Cell Technology and Regenerative Medicine Towards Application

Kakon Nag, Nihad Adnan, Koichi Kutsuzawa and Toshihiro Akaike

Additional information is available at the end of the chapter

http://dx.doi.org/10.5772/58287

1. Introduction

'Stem cell' – the term was first coined by Russian histologist Alexander Maksimov in 1908 to herald the existence of special cells those have capacity to generate blood cell. Stem cells are the core materials of regenerative medicine and tissue engineering. Although there are multiple types of stem cells available based on their origin and functionality; however, scientifically they can be classified into four well-defined classes– (1) embryonic stem cell (ESC), (2) adult stem cells (ASC) for example, muscle satellite cells are muscle-specific adult stem cell, (3) induced pluripotent stem cell (iPSC), and (4) pathological stem cells (PSC) for example, cancer stem cells (CSC) [1]. Out of these 4 types, ESC and ASCs are true physiological stem cells, iPSCs are engineered stem cells and PSCs are conditional stem cells. Among them, ESC and iPSC are being considered true pluripotent stem cells, which have the capacity for unlimited self-renewal and differentiation into all the specialized cell types of the body. Therefore these cells have been considered the most favorable cells for using in regenerative medicine and tissue engineering [2,3,4,5,6,7,8].

Stem cells need a special environment for their survival, maintenance and growth. During the early stage of establishing the culture methodologies for stem cells, it was realized that they need support from other cells for example, mouse embryonic fibroblast (MEF). Co-culture methodology with gamma-irradiated MEF cells used as feeder-cells and enriched culture media with fetal bovine serum (FBS) were successfully utilized for establishing *in vitro* stem cell culture [9,10]. However, using a second non-related cell type (although growth restricted) is not suitable for differentiation studies – particularly, for 3D cell culture. Later, the MEF layer was successfully removed from the culture system by introducing MEF-conditioned media (MEF-CM) that has made the protocol more suitable for experimentation targeting regenerative medicine but not up to the desired standard due to the presence of xenogeneic agents in the system [11,12,13]. MEF-CM is enriched cell culture media with MEF-secreted molecules that functions as a depot for the necessary cytokines for the healthy maintenance of stem cells. However MEF-CM alone were not adequate to upkeep ESC and iPSC survival and growth thereby suggesting that MEF cells are not only providing necessary nutrients and cytokines, in addition they are also backing as physicochemical supports through the ECM to these cells. However, technically it remains elusive to point out the essential factors, required to maintain stem cell culture, present in the MEF-CM due to the inconsistency in expression and secretion of biological factors between experiments and batches. Moreover, it has been shown that not only proliferation of these cells but the secretion of necessary biomolecules and deposition of ECM components were also directly related to the gamma-irradiation [11,12,13]. Such factors directly influence properties of stem cells in culture, and instigate restriction for application of relevant protocols for regenerative medicine and tissue engineering. Therefore suitable cell-recognizable biomaterials are highly desired to overcome the dependency of cell-based basal supports for stem cell culture.

Matrigel was one of the first biomaterials that was effectively applied as plate-coating materials for *in vitro* culture of human ESC and iPSC with the aid of MEF-CM as culture medium [14,15]. This was a significant advancement in stem cell technology to make stem cells free from undesirable feeder-layer cells. Matrigel is a product from decellularization of Engelberth-Holm-Swarm (EHS) mouse sarcoma cells, and a cocktail of laminin, collagen IV, entactin, heparin sulfate proteoglycans, and known and unknown growth factors with variable compositions [16,17,18,19]. It closely resembles the embryonic basement membrane in consistency and activity as well as providing a biologically functional complex [17,19]. However, Matrigel is not a defined material with high purity and incorporated with substantial lot to lot variation in constituents both in qualitative and quantitative measures. It has also been reported contaminated with Lactate Dehydrogenage Elevating Virus, and has raised additional concerns for safe application of this material in stem cell culture [20]. Such kinds of issues are strongly demanding a more defined culture condition under good manufacturing practice (GMP) for safe application of stem cell protocols or methodologies if the ultimate objective is to employ stem cells in regenerative medicine or tissue engineering.

The individual components of Matrigel provide specific functional queues to ESCs and iPSCs. For example, ESC exhibits normal growth when cultured on laminin-coated plate, which was not observed on either fibronectin- or collagen IV-coated surface [21,22,23,24]. It was also reported that specific laminin isoforms have distinctive effects on stem cells; for instance, laminin-111, -332, -511 support adhesion and proliferation of stem cells but isoforms -211 and -411 of laminin do not [22]. The information suggested that designing a defined matrix for stem cell culture requires special biomaterials that can deliver concurrent supports for cell adhesion, proliferation and differentiation. In fact, effective stem cell culture condition with high pluripotency was occasionally achieved in spite of introducing several synthetic and semi-synthetic biomaterials alone or as a blend as cell-culture substrate, and therefore, designing such a biomaterial remains a challenging but ultimately rewarding task.

Pioneering work from our laboratory introduced Fc-chimeric protein in stem cell technology approximately a decade ago, and over the years we and others have established multiple Fc-chimeric proteins as significantly favorable cell-recognizable biomaterials in stem cell technology. These works with varieties of Fc-chimeric proteins spanning from ECM component protein [for example, E-cadherin (ECad)] to cytokine [for example, hepatocyte growth factor (HGF)] have shown tremendous potential to overcome the major barriers in stem cell technology, namely defined condition for stem cell culture, selective differentiation to the target lineages, convenient purification of the desired cells etc., for the application of stem cell technology targeting to regenerative medicine. In this article we will focus on ECad-Fc and NCad-Fc chimeric proteins as novel cell-recognizable biomaterials in stem cell technology towards application in regenerative medicine.

2. Rationale for using protein as biomaterials

An ideal chemically defined xenogeneic-agent free stem cell culture system might be consists of chemically known matrix for plate coating that would provide structural basal support to the stem cells and defined media that is supplemented with highly pure recombinant proteins as functional cytokines. The system should essentially be free from serum or feeder-cells or any other animal products. Even though it is very demanding however, designing and preparing a completely defined stem cell culture system is highly challenging. One worthwhile goal is to design a defined plate-coating material that can successfully replace Matrigel. Since stem cells are essentially dependent on cell-cell or cell-surface interaction for survival, which are mainly mediated by extracellular matrix protein (ECM), a cell-recognizable biomaterial should preferably mimic ECM protein(s).

Such kind of biomaterials can either be employed as a scaffolding molecule that may provide structural support of the growing cells, or as functional effector molecules that can target cellular signal recognition machineries like cell surface receptors or channels to trigger or maintain signaling cascades necessary for survival, proliferation, and differentiation of experimental cells [25]. To act as an artificial ECM the biomaterial under consideration should mimic the physicochemical and biological properties of native components of ECM to facilitate targeted functionalities of cell for example, adhesion, proliferation, differentiation, etc [26]. Similarly, the candidate effector molecules should have physicochemical signature of the comparable native molecules for recognition as functional substrate to endogenous receptors or channels of experimental cells. Synthetic biomaterials have limitations for providing perfect biochemical structural motif for effective recognition by the cellular recognition machineries to execute necessary cellular function, and therefore are generally not efficient enough for practical applications for *in vivo* condition. Moreover, many of these synthetic biomaterials are not biologically compatible at a desired level and may generate pathophysiological complications in the long term in the body.

Proteins are native elements of cells and natural ECM scaffolds [27] and therefore recombinant proteins could be one of the best candidates to design superior biomaterial for application in regenerative medicine and tissue engineering. Recent progress in biochemistry, molecular biology, bioinformatics, and engineering provides the prospect of expressing and purifying desired recombinant protein with high yield (g/L is achievable) in large scale [28], which can eventually be applied (directly or with modification) as novel, simplified, and bio-active macromolecules in regenerative medicine and tissue engineering [29,30,31]. Such proteins can be generated from a genetic template by natural cellular read-out process namely, DNA>RNA>protein that ensures excellent uniformity and reproducibility of the designed biomaterial depending on cellular conditions, where the production is executed. The native biological production process confirms high degree of reproducibility, which is not realistic by traditional chemosynthetic or mechanosynthetic processes. On the contrary, protein science has its own negative issues for example, highly efficient expression system for the desired protein, convenient purification of the target protein, proper folding of the purified protein, stability of the functional protein, mode of application of experimental protein etc. Chimeric protein technology has long been considered one of the potential methodologies to overcome many of these issues including higher productivity, better stability, and efficient purification of a target protein for bulk scale. Fc-chimeric protein is one such engineered protein that was introduced in 1989, and has been showing great promise for comparatively convenient production efficiency of chimeric protein with functional integrity and long-term stability, and therefore successful applcation in diverse fields of biomedical sciences [32,33,34]. An illustration of Fc-chimeric protein is shown in Fig. 1 with ECad-Fc as a model.

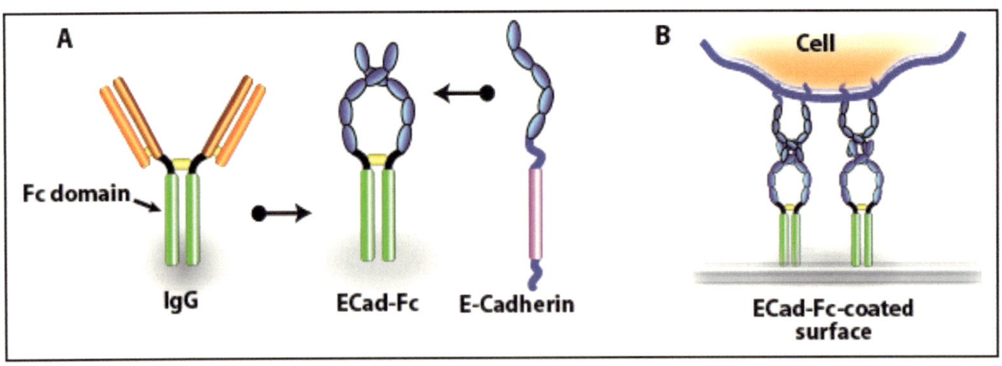

Figure 1. Schematics of Fc-chimeric protein, and its molecular function. (A) Functional domain of target protein is fused as N-terminal with the Fc domain of IgG. ECad is shown here as an example. (B) Plasma-membrane localized ECad dimer can interact with apposing ECad dimer and form high affinity binding that makes cell-cell and cell-surface adhesion.

3. Cadherins in cell biology

The cadherins is a large family of single transmembrane proteins with more than 100 members. Out of these we will be focusing on epithelial cadherin (ECad) and neural cadherin (NCad) in this report. They are the member of classical cadherin family, and both of them are glycosylated in their extracellular domain. They have the ability to function as adhesion molecules for the relevant protein-expressing cells. Generally cadherin forms homophilic dimer, and the dimeric forms of cadherins take part in Ca^{+2}-dependent coupling from apposing cells that mediates cell-cell adhesion. These single transmembrane-domain plasma membrane-resident proteins are not only necessary for cell-cell adhesion but also involved in indispensible signaling cascades, which are critical for the development-to-homeostasis-to-demise of cells and organisms.

The extracellular N terminal region of ECad consists of 5 structural domains, which are the signature motifs for ECad and are responsible for the homophilic binding between two neighboring as well as apposing molecules, while the C-terminal intracellular region of ECad interacts with several intracellular proteins such as β-catenin/Armadillo and p-120 catenin [35,36,37]. The p-120 catenin is associated with the targeted transport and stabilization of the adhesion complexes on the plasma membrane. Beside, β-catenin interacts with α-catenin, which in turn initiates actin filament formation *via* interaction with formin at the adherens junction [38,39,40,41,42]. However, how cadherin-catenin complexes are connected with cytoskeletal components *e.g.*, actin is not clearly known.

ECad has been shown linked with many early-to-late developmental and differentiation processes *in vivo* and *in vitro* systems including ESCs, MSCs, iPSCs, and whole embryo [43,44,45,46,47]. ECad knock out mouse was reported embryonic lethal [48,49], which is a direct evidence of its critical importance in stem cell biology and regenerative medicine. Our lab first envisioned the application of ECad as a novel cell-recognizable biomaterial little over a decade ago while Nagaoka *et al.* endeavored to improve the differentiation and maturation efficiency of hepatocyte in an *in vitro* system [50]. The idea was conceived from the fact that Fc domain of IgG can bind directionally with an appropriate surface *via* hydrophobic interaction, and the fused protein stretches out directionally to offer interaction with a suitable partner [51]. At that period, several reports suggested that ECad is indispensable for tissue morphogenesis, and is also required for maintenance of matured tissues. Awata et al. showed that ECad-mediated cell-cell interaction is necessary for hepatocytes to maintain their differentiated phenotypes by forming 3D spheroid structure, or multi-layer cell aggregates [52]. Further it was reported that high cell density culture of fetal liver cells [53,54], which most likely is an ECad-dependent characteristics, enhanced hepatocyte maturation in culture. These findings suggested that cell-cell interaction may directly influence hepatocyte maturation as well as maintenance of differentiated phenotypes. There was, however, no substantial information regarding the role of ECad in the relevant processes, and to reveal the answer it was essential to have a suitable tool or methodology that can expedite cell-cell interaction analysis in a controlled manner. ECad-Fc was designed and deployed as a novel biomaterial in the regenerative medicine field to address this issue; after a decade, it has been proven to be a suitable material for stem cell technology and regenerative medicine.

4. ECad-Fc as a cell-recognizable biomaterial

As a biomaterial, ECad-Fc was first applied as plate-coating materials for hepatocyte differentiation experiments [50]. It was observed that differentiated hepatocytes can efficiently adhere with the cell culture plate coated with ECad-Fc. The adhered cells demonstrated comparable molecular characteristics e.g., low DNA synthesizing activity and maintenance of tryptophan oxygenase (TO) expression like those of spheroid-form hepatocytes. As well, the hepatocyte cultured on ECad-Fc-coated plate supported the differentiation of hepatocytes in culture. These results suggested important roles of ECad-Fc matrix for the maintenance of differentiating hepatocytes. This was the first report of ECad-mediated matrix dependability, as a biomaterial, for any cell type in regenerative medicine. After a while, Nagaoka *et al.* published the landmark report regarding the application of ECad-Fc cell-cooking plate (since target cell can be obtained on such type of biomaterial-coated plate without additional cell purification method therefore named so) as a defined matrix for successful maintenance of murine stem cells without any feeder layer in 2006 [55]. This report signified the alluring potential of ECad-Fc as a biomaterial for practical application in stem cell technology and regenerative medicine.

Xenogeneic-agent free stem cell culture method is extremely critical if the objective of the relevant protocol is to apply the relevant products in regenerative medicine. Since MEF secrets many unidentified molecules, which are potential xenogeneic elements for human subject therefore feeder-cell-based early methodologies are not considerable for applying in regenerative medicine. Matrigel is also produced from mouse carcinoma tissue and ill-defined therefore causing serious known and unknown hazards of xenogeneic contamination in experimentations. An immunogenic sialic acid (NeuGc) has been identified in a co-culture experiment for human ESCs applying MEF and animal derivatives as serum replacement [24,56]. This is specifically worrying as such kind of non-human sialic acid can initiate immunogenic processes in human triggering complete graft rejection and consequential complexities. Non-human animal-derived products also can be a possible cause for mycoplasma contamination, which can directly infect the cells in culture and either damage them totally or can change their properties, and thereby directly or indirectly initiate complicacies for regenerative medicine protocols. Human feeder-cells and serum have been recommended for culturing human ESCs to evade xenogeneic compound in experimental system for regenerative medicine. However, this is associated with a high risk of microbial contamination, for example retroviral components, and hence are not as suitable for *in vivo* application. Therefore it is a prime importance to establish completely defined human stem cell culture system for safe application of relevant products in regenerative medicine.

5. ECad-Fc is a unique defined matrix for ESC and MSC

The study of Nagaoka *et al.* [55] revealed that murine ESCs can maintain their pluripotency on ECad-Fc-coated surface for extended culture periods (Fig. 2). Cells cultured on such type of substratum were later successfully used to generate germline-competent chimeric mouse [57]. Consistent with the findings, a separate study using mouse mesenchymal cell lines STO and NIH3T3 stably expressed with ECad as feeder-cell showed higher level of stem cell marker expression with standard colony-forming phenotype compare to the cells cultured on normal MEF-feeder-cell layer [58]. A number of feeder-free culture methods for ESCs have been reported where ESCs grow with their standard tightly-bound colony phenotype [4,11,13,22,24,56,59]. This type of tight colony formation generates heterogeneous cell population within a colony, which potentially affects homogenous accessibility of cytokines to these cells as well as creates heterogeneous niches. As a result stem cells in a colony differentiate heterogeneously and produce various kinds of cells as contamination with the desired type of cells, a major drawback that regenerative medicine has to overcome. In this respect, ECad-Fc matrix drives murine stem cells out of the colony to form a normal monolayer of cells, where stem cell resides as single cell condition [55]. This is a ground breaking technology that provides an exciting solution for overcoming the inherent colony forming phenotype-linked cellular heterogeneity. Biochemical analyses revealed that these cells bear all the signatures of pluripotent stem cells, and can form all three germ layers in a teratoma forming assay, and as mentioned earlier can generate germline-competent chimeric mouse. Additionally, they require lower amounts of LIF for maintenance of pluripotency, reducing costs related to ESCs culture. The monolayer-type single cell ESCs was also associated with higher proliferation ability and greater transfection efficiency compared to the colony-forming cells cultured on other substratum. Such improved proliferation ability could be extremely helpful for quick amplification of iPSCs on ECad-Fc substratum, which could mean shorter waiting periods for patients to receive cell therapy. The higher transfection efficiency of stem cells on ECad-Fc cooking plate could be exploited for targeted delivery of desired extracellular cargo for example, transgene products or drug molecules, into these cells for better outcomes.

This type of cooking-plate technology, where ECad-Fc provides basal support to the cells, and other immobilized factors for example, LIF-Fc [57] which satisfy specific needs, can be very advantageous for (1) ensuring undifferentiated state of stem cell in culture, (2) cost reduction associated with cytokines, and (3) hassle-free working condition without the necessity of regular media change, which is a standard time-consuming practice for stem cell culture.

The single-cell phenotype seen for ESCs was also observed for other stem cells for example, mouse embryonal carcinoma cells F9 and P19 but not for differentiated cells for example, NMuMG mouse mammary gland cells, MDCK kidney epithelial cells and isolated mouse primary hepatocytes [60]. This result indicated that ECad-Fc-mediated cellular migratory behaviors are most likely specific for embryonic stem cells. Reportable that ECad-facilitated cell-cell adhesion is often rearranged during initial stages of embryogenesis to control cell migration, cell sorting, and tissue function, which is suggesting a close cooperativity of stem cell maintenance, proliferation, and differentiation with ECad [39,48,49,61,62]. However, there

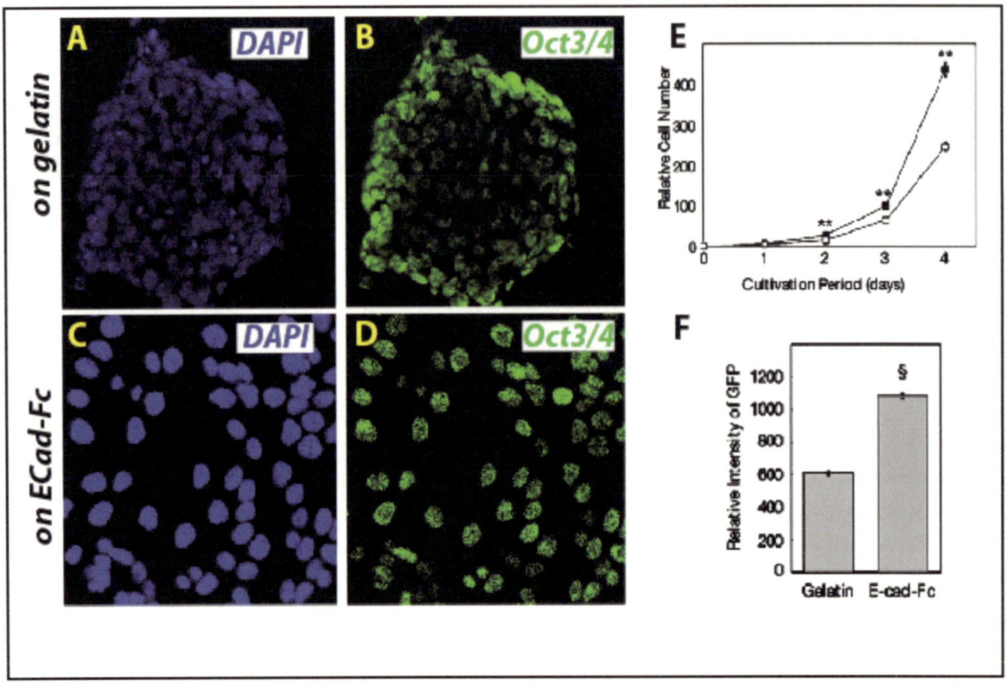

Figure 2. ECad-Fc is a defined matrix for culturing monolayer of iPS cells. Mouse EB3 cells were successfully cultured on ECad-Fc-coated surface that showed monolayer phenotype (C and D) compare with compact colony phenotype (A and B) for general protocol, which was significantly advantageous for faster growth (E), and higher transfection eficiency (F).

is no such suitable system to explore the necessary signaling pathways to address these questions. Nevertheless, since ESC does not form colony on ECad-Fc cell-cooking plate therefore this can be a perfect tool for obtaining single cell model system of stem cells to investigate relevant signaling pathways necessary for stem cell maintenance, proliferation, and differentiation. Our recent study successfully exploited this single-cell phenotype for monitoring cell cycle properties of stem cells on cell-cooking plate (unpublished), indicating the importance of this system for cell biology experiments designed to reveal their individual characteristics. The findings could be invaluable for regulating stem cells for desired application in regenerative medicine.

Most of the stem cell innovations, comprising generation of ESCs and iPSCs, were primarily established in mouse model, and then applied in human models. Similarly, ECad-Fc cell-cooking plate technology was first developed and established for murine stem cells [55,57]. Thereafter, ECad-Fc cooking-plate was successfully applied for human ESC culture following similar methodologies with additional consideration for mild enzymatic treatment during the cell dissociation and seeding steps [56]. A strong protease cocktail Accutase (Millipore) was used for murine ESC culture; however, Accutase treatment was found detrimental to human ESCs, which was recuperated by using enzyme-free proprietary preparation named, Cell Dissociation Buffer (Life Technologies). It is reportable that the human ESCs were cultured on

ECad-Fc cooking plate with a completely defined media named mTeSR1 (Stemcell Technologies), and that made the culture method completely defined and xenogeneic-agent free, which is a significant achievement in regenerative medicine. The stem cells cultured on ECad-Fc cooking-plate were practically identical to those cultured on Matrigel-coated plate including cell morphology, proliferation rate, preservation of undifferentiated phenotype, and ability of differentiation into multiple cell types in embryoid bodies as well as in teratoma assay [56]. Interestingly, contrasting with the single-cell phenotype for mouse ESCs, human ESCs produced normal colony forming phenotype on ECad-Fc cooking-plate. The mechanism underlying the difference for this observation was not completely understood though.

Human and mouse ESCs have been shown to demonstrate significant disparities in expression of cell surface markers, transcription factors, cytokines, and proteins in them. The difference was evidently recognized by the fact that mouse ESC can be maintained in undifferentiated state with the addition of LIF devoid of feeder-cell but human ESC cannot [14]. It has been shown that the inhibition of Rho-ROCK signaling pathway generates cell scattering in human ESCs suggesting direct connection between cell scattering and signaling pathways [63]. While both mouse and human ESCs express ECad, however, it appears there are diverse additional factors involved to define ECad-mediated activities in these cells and additional investigations are required to reveal the complete molecular circuitry associated to this phenomenon.

MSC is a type of ASCs, and can be collected from donor by satisfying approved ethical issues. These cells have been considered as potential starting materials for regenerative medicine and tissue engineering. They must be expanded *in vitro* before dispensing for specific applications to accomplish anticipated therapeutic effects. MSCs also need xenogeneic agent-free culture method for maintaining their differentiation potency over the culture period. ECad-Fc cooking-plate technology was effectively applied for this reason as well [43]. The cultured MSCs on human ECad-Fc (hECad-Fc) matrix exhibited superior attachment on culture plate compare with standard tissue culture plate and gelatin-coated plate. The MSCs cultured on hECad-Fc showed comparable level of CD 105 and significantly greater level of β-catenin and ECad expression. It has been reported that β-catenin enhances the activity of Oct-4, which is one of the principal Yamanaka factors that plays critical function during the regulation of self-renewal of ESC [45,64], on conjecture it can be suggested that MSCs maintained on ECad-FC cooking-plate might preserve superior stem-ness compare to the MSCs maintained on tissue culture-treated plate and gelatin-coated plate, and therefore possess greater applicability for regenerative medicine.

6. ECad-Fc in directed differentiation and *in-situ* cell sorting of stem cell

Targeted differentiation of stem cells and enrichment of desired cell for example, hepatocytes, from the pool of differentiated cells are very important steps towards use of the cells for regenerative medicine. Functionally matured hepatocytes derived from stem cells can be a potential remedy for various hepatic diseases. There have been several hepatic differentiation protocols reported from ESCs using orthodox techniques including embryonic body (EB) formation, and clustered colony formation on gelatin- or feeder-cell-coated plates [52,54]. However, these protocols come with many drawbacks, for example, heterogeneous cell population, spontaneous differentiation, xenogeneic contamination, inefficient conversion to hepatocytes, requirement for enrichment of target cell population etc. Our group has effectively applied ECad-Fc as a cell-recognizable plate-coating materials that facilitated good quality mouse ESCs in culture with superior proliferative activities and single-cell phenotype. Similarly, the cell-recognition property of such Cadherin-Fc chimeric protein was exploited for the possibility of facilitated differentiation of ESCs to specific cells for example, hepatocytes and neural cells [29,30,50,65]. Remarkably, ECad-Fc substratum favored progressive differentiation of ESCs to cells with features of definitive endoderm, hepatic progenitor cells, and finally phenotypical as well as functional hepatocytes-like cells [30,50]. The ECad-Fc-coated substratum stimulated selective hepatocyte differentiation in association with ectopic hepatocyte-producing cocktail resulting around 55% hepatic endoderm cells devoid of neuroectoderm and mesoderm markers [30]. High level of (approximately 98%) ECad and developing-hepatocyte marker α-fetoprotein (FTP) were co-expressed in these cells. Since these differentiating hepatocytes express high level of ECad on the plasma membrane therefore ECad-Fc was employed for on-site one-step enrichment of *de novo* hepatocyte-like cells. Practically, 92% albumin expressing cells were successfully harvested on ECad-Fc cooking-plate without any harsh enzymatic treatment or mechanical cell sorting, which are usually detrimental for cells [30]. Therefore the technology can be successfully applied for quick and stress-free cell purification, which will be useful in regenerative medicine.

The enhanced differentiation and cell-recognizable properties were also observed with ECad-Fc and NCad-Fc-based mixed biomaterial cooking-plate for neural cells [65], and is discussed in detail under NCad-Fc section. Such kind of ECad-Fc and NCad-Fc hybrid cooking-plate can be applied for either generation of large number of homogeneous cell population, which can be applied for therapeutic evaluation, or for analyzing the signaling pathways related to nerve generation at a single cell level.

7. ECad-Fc is a superior matrix for iPSC

iPSCs are commonly derived from somatic cells by ectopic and forced expression of common transcription factors Oct4, Sox2, and Nanog along with protocol-dependent treatments with cocktails of some other transcription factors, and even miRNA or small molecules [10,66,67,68,69,70,71,72,73]. Despite the existence of many protocols for generating iPSCs, the required time and efficiency of iPSC generation is still not practical for application of the technology to a mass scale. As per recent published information, depending on protocol, it may take somewhere between 2~4 weeks to get a 1% conversion of cells to iPSCs. During the reprograming process, starting cells experience mesenchymal-to-epithelial transitions (METs) as a natural requirement [74]. This fact was further proved by the findings that MET happens during the initial stage of reprograming process [71,74,75]. Recent evidence further suggested significant functional roles of ECad and other cell adhesion molecules in METs.

ECad interacts with cytoskeletal components *via* various intracellular molecules for example, α-catenin, β-catenin, and p-120 [38]. ECad-mediated signaling was found associated with cytoskeletal remodeling processes through Rho activation [41,63,76]. ECad has been established as an essential factor for maintaining typical colony-forming phenotype of ESCs and iPSCs. Recent studies, remarkably, revealed that forced expression of ECad can significantly enhance the effectiveness of relevant iPSCs-generation protocol [45]. A separate study revealed that ECad expression was enhanced upon treatment with small molecules resulting in enhanced efficiency for the relevant iPSC-generation protocol [77]. This enhanced productivity for iPSCs was successfully reproduced by the application of N terminal extracellular domains of ECad, which suggested that the phenomenon is mainly mediated by the extracellular functional domains of this protein [77]. Most importantly, ECad was sufficient to generate iPSCs with only three Yamanaka factors –KLF4, SOX2, and c-MYC from murine fibroblasts without OCT4 [45]. This study indicated that the spatial and mechanical input exerted by ECad has a critical role in driving cell fate. However, it is not clearly understood how ECad can compensate for OCT4. Since many studies showed that where it was possible to skip other factors of Yamanaka-cocktail for reprograming of somatic cells to iPSCs but OCT4 was hardly indispensable [10,70,78], further studies are warranted to determine the underlying mechanism. One potential explanation might be that ECad and KLF4 together initiated an early MET process of the experimental cells, and then SOX2 and KLF4 operated co-operatively to propel pluripotency genes to induce initiation of reprogramming [74]. The hypothesis is favored by the fact that cells those already express ECad, for example keratinocytes, can be reprogrammed more effectively and quicker because the MET process is not required [71]. Since the extracellular domain of ECad is adequate to produce ECad-mediated influences related to the reprogramming of somatic cells to iPSCs we have therefore assumed that ECad-Fc could significantly enhance the reprogramming efficiency. Our preliminary observation suggested that indeed co-transfection of ECad-Fc-expressing plasmid with Yamanaka factors enhanced reprograming efficiency of mouse fibroblast (unpublished). Enhanced reprogramming efficiency was further witnessed while the Yamanaka-cocktail-transfected starting cells were cultured on ECad-Fc-coated plate compare to gelatin-coated plate. However, further experiments are necessary for providing detail quantitative and qualitative information for these observations. Nonetheless this finding is highly promising regarding enhanced and efficient generation of iPSCs using a biomaterial as substratum.

The protocols for generating ESCs or iPSCs as well as differentiation to target cells from these cells require cell isolation step either by mechanical process or in combination with enzymatic treatment [79]. These types of methodologies require skilled labor, specialized instrumentation, additional time and cost, and distinct morphologic and phenotypic features. Several protocols have been described recently for enzyme-selective passage of specific cells; however, they are not globally applicable and very often appeared with unwanted cells. Enzymatic treatment also caused karyotypic anomalies compared with manual passaging [66,80]. FACS protocol has been applied for cell sorting based on surface marker recognition. However, relevant protocols need enzymatic treatment, application of foreign molecules, and mechanical processes involving severe stress on experimental cells [81,82], which are highly unfavorable for cells. ECad-Fc cooking-plate, advantageously, neither needs any kind of mechanical sorting nor any harsh chemical or enzymatic treatment. The experimental cells can selectively and strongly make homophilic binding with ECad-Fc matrix in a Ca^{+2}-dependent manner subjected to the differential expression pattern of ECad in them during the transformation process. The cells with no or low level expression of ECad cannot and does not firmly bind with ECad-Fc substratum and can be washed off with suitable buffer thus offering a unique, robust, and stress-free cell enrichment system. Such a protocol ensures quicker, cheaper and convenient cell enrichment system for *in vitro* culture without risk of additional contamination and cellular alteration, and therefore, is highly advantageous for application in regenerative medicine and tissue engineering to achieve desired therapeutic effect with minimal adverse consequences.

8. NCad in cell biology

N-cadherin (NCad) or neural cadherin is also known as Cadherin-2, which is encoded in human by *CADH2* gene [83,84]. Like ECad, it is also a cell-cell adhesion molecule composed of five extracellular cadherin domains, a transmembrane domain and a highly conserved cytoplasmic region. NCad can exist either as strand dimers or in an alternate monomeric form [85]. NCad typically forms homotypic homophilic interactions between two neighbouring cells for example, Sertoli cells and spermatides, and also heterotypic homophilic and heterophilic interactions, such as interaction between N- and R-cadherin in transfected L cells [86]; such interactions are Ca^{+2} dependent [87], and can be reversed by withdrawing Ca^{+2} from the system.

During embryogenesis cells undergo an epithelial-mesenchymal transition (EMT) initiating upregulation of NCad and the downregulation of ECad in the mesoderm [88]. It has been suggested that NCad expression is essential for morphogenesis of the mesodermal germ layer during gastrulation [89]. NCad expression pattern has been found complementary to that of ECad in epidermal ectoderm [88,90,91]. NCad expression has been detected in mesoderm and notochord in the early phase of embryonic development, which is later also evident in neural tissue, lens placode [92], some epithelial tissues, myocardium of heart [93], epiblast of skeletal muscle [94], endothelial cells, osteoblasts, mesothelium, limb cartilage, and primordial germ cells [95,96].

NCad is found to be present in the early hematopoietic progenitor CD34+CD19+ cells, and it was proposed that NCad plays critical role for the hematopoietic cell differentiation as well as the early retention of this subpopulation in bone marrow [97]. During skeletal muscle formation mesodermal precursors exit from the cell cycle, and differentiate into myoblasts that terminally differentiates into multinucleate myofibers [98]. Cell cycle arrest and the expression of skeletal muscle–specific genes are the critical checkpoints for this developmental process [99]. All the epiblast cells undergoing skeletal myogenesis express the skeletal muscle-specific transcription factor MyoD, among them only the cells expressing NCad but not ECad can differentiate into skeletal muscle [94]. NCad function-perturbing antibodies showed that it plays a significant role in interaction between myoblasts in myotube formation and in myofibrillogenesis [100,101,102]. NCad is also found to be involved in myoblast migration in limb bud [103].

Cartilage is formed from the vertebrate embryonic limb by a highly synchronized and systematic event of cell commitment, condensation and chondrogenic differentiation of mesenchymal cells to chondrogens, and by the production of cartilaginous matrix. SOX9, an essential transcription factor for chondrocyte differentiation and cartilage formation, binds to the SOX9-binding motif in NCad promoter [104] that facilitates expressing of NCad gene products to play necessary roles in cellular condensation [105]. Prolonged expression of NCad due to the missexpression of wnt7a stabilizes NCad-mediated cell-cell adhesion resulting inhibition of chondrogenesis from mesenchymal chondrogenic culture [106]. The level of NCad mRNA was found increases during osteoblast differentiation and decreased during adipogenic differentiation thus suggesting their involvement in relevant differentiation processes [107]. NCad expression is increased in osteoblasts by BMP-2, FGF-2 and phorbol ester (e.g., PMA) in PKC-dependent manner, whereas factors like TNFα and IL-1 reduce the expression of NCad [108].

Migratory cell populations, also known as neural crest cells, are pluripotent cells those originate from dorsal part of neural tube and play important roles in embryonic development and pathophysiological conditions. These cells express NCad when they are associated with neural tube; however, NCad expression is down-regulated after EMT process and the relevant cells started to migrate over long distance, and finally transform into different types of tissues and cell populations, such as peripheral nervous system, cartilage, bone and melanocytes. Slug plays here important roles in down-regulating NCad that leads to a loss of cell-cell adhesion and allowing the cells to migrate. The dorso-ventral migratory cells re-express NCad during dorsal root and sympathetic ganglia developmental steps and promotes cell aggregation; thereafter, only dermal melanocytes express NCad [109,110]. This observation is suggesting critical involvement of NCad in the development of relevant tissues.

Several proteins can interact with NCad *via* intracellular and extracellular domains and influence subsequent signaling pathways. The functions of NCad in controlling neurite outgrowth, synaptic plasticity and guidance in synapse formation have been proposed [111]. These functions may involve interaction with other membrane bound molecules, such as fibroblast growth factor receptor (FGFR), which was confirmed by blocking the FGFR by pharmacological inhibitor [112]. NCad directly interacts with FGFR *via* HAV epitope of FGFR with IDPVNGQ epitope of EC4 of NCad [112], and this interaction between NCad and FGFR can be of both ligand dependent and independent [113] suggesting wider cooperative functional significance of this duo in relevant development and physiology.

EMT of squamous epithelial cells ectopically expressed specific amino acid sequences of EC4 of NCad induces motility. The cell motility behavior and adhesion is independent to each other, as antibody against the aforementioned relevant amino acid sequence of NCad inhibits cell motility but the cell-cell adhesion phenomena was uninterrupted [114]. The influence of NCad mediated cell migration is cell type specific, as it was found that NCad can inhibit LM8 mouse osteosarcoma cell migration but it did not have any significant effect on the movement of MDA-MB-435 cells [115]. The cytoplasmic domain of NCad form complexes with various types of molecules, such as p120, β-catenin, α-catenin and GAP-43, and regulate various cytoskeletal dynamics. All of these interactions are critically involved in tissue-to-animal development, morphogenesis and maturation, and is suggesting the possibility of exploiting this gene product for regenerative medicine.

9. NCad-Fc as biomaterial in regenerative medicine

NCad-Fc was introduced by Lambert et al. in 2000, and the study revealed that NCad-Fc not only induced the recruitment of NCad on the plasma membrane but also other components of the cadherin/catenin complex. This work for the first time demonstrated that NCad-Fc can mimic natural cell-cell contact formation and signal transduction [116]. Pioneering work from our lab has introduced NCad-Fc as cell-coating biomaterials for stem cell culture. NCad-Fc protein was collected from 'pRC-NCFC' plasmid, which was constructed by inserting the N terminal extracellular domain of mouse NCad into pRC/CMV (Invitrogen) plasmid [29]. The expression and purification methodologies of NCad-Fc are similar like ECad-Fc and have been described in details in relevant publications [50,55,117]. Over recent years our laboratory work revealed significant advantages of NCad-Fc in neural differentiation from stem cells. Early work was performed with mouse embryonic carcinoma cell P19 and neural stem cell MEB5 because of their easy management over the ESCs. It was observed that culturing these cell lines on NCad-Fc substratum can maintain the undifferentiated state and scattering morphology compare with other control substratum such as gelatin, fibronectin, laminin or poly-L-ornithine. P19 and MEB5 cells were differentiated effectively to neural lineage on this defined matrix in presence of retinoic acid supplemented with insulin-transferrin-selenium commercial preparation (ITS, Invitrogen). Interestingly, P19 cells showed higher level of *Neurog1* expression on NCad-Fc-coated surface compare with gelatin-coated surface. Additionally, MEB5 differentiated on NCad-Fc matrix, compared to fibronectin-coated surface, showed complete neuronal differentiation phenomena and significantly higher expression levels of neural markers, such as *Neurog1* and *MAP2*. These results clearly suggested the superiority of NCad-Fc substratum over the other experimental substratum for neuronal differentiation process.

Later, the findings were extrapolated to MEF-dependent mouse embryonic stem cell ST1 and mouse iPSCs to evaluate whether the effect is restricted to specific pre-committed cell lines or it is globally applicable [65]. Since during EMT conversion ECad is downregulated and NCad is upregulated therefore a hybrid matrix of ECad-Fc and N-Cad-Fc was designed to exploit the stage-specific cadherin switching phenomenon. The concept was that, initially the ESCs and iPSCs would bind to ECad-Fc through cell-resident ECad, however, during and after neuroectoderm formation cadherin switching will cater for cellular NCad in place of ECad that would bind to NCad-Fc. The cadherin switching was experimentally confirmed in house during neural differentiation protocol (Fig. 3A), where Dkk-1, a Wnt signaling pathway antagonist, and LeftyA, a Nodal signaling pathway antagonist were used for triggering neural differentiation. Specific markers for primitive ectoderm, primitive neural stem cells, neural stem and progenitor cells were checked. Along with, promisingly, the efficiency of neural progenitor differentiation from mouse ESCs on cadherin-Fc chimeric matrix was significantly higher compare to the cells cultured on other standard substratum as evaluated by the higher

http://dx.doi.org/10.5772/58287

level of expression of neural progenitor marker Nestin gene products. Furthermore, the differentiated cells exhibited greater levels expression of βIII-tubulin (Tuj1) (Fig. 3B), microtubule associated protein 2 (MAP2), Pax6, and tyrosine hydroxylase but not GFAP, which is a marker of glial cell, signifying the presence of a lineage confined to neural cells.

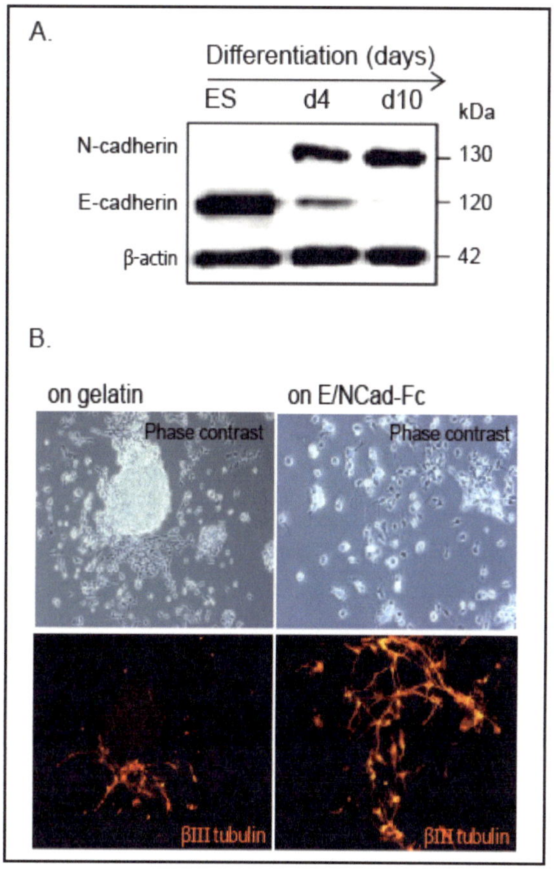

Figure 3. NCad-Fc, and ECad-Fc promote directed differentiation of target lineage from iPSCs. (A) Western blot data revealed ECad to NCad switching occurs during neuronal differentiation. The expression level was normalized using house-keeping gene, β-actin. (B) βIII-tubulin expression was significantly higher on E/NCad-Fc matrix compared to gelatin.

Culturing of ESCs and iPSCs on ECad-Fc and NCad-Fc hybrid substratum not only developed scattered cell morphology as reported for ECad-Fc substratum but higher cell proliferation rate and enhanced differentiation efficiency were also noted. Along with these phenomena significant higher degree of homogeneity and enhanced differentiation efficiency were also observed, which is a remarkable advantage for harvesting target neuronal cells from *in vitro* system that can later be applied for regenerative medicine protocol. Although EB-based protocols are being relatively well-practiced for neural differentiation, however, the inconsistency of the embryoid body (EB) size and shape, and the asynchronous distribution of growth factors throughout the EBs give rise to heterogeneous products. Besides, monitoring cell morphology during differentiation process for EB-based differentiation protocols is inconvenient. Culturing ESCs or iPSCs in scattered single cell condition, on the contrary, can effectively overcome these issues. Interestingly, this blend of Cadherin-Fc matrices maintained a complete homogeneous cell population for murine ESCs and iPSCs for several passages. Highly homogeneous population of primitive ectoderm and neural progenitor cells were routinely generated on such a hybrid-type cooking-plate [65]. Enriched population of neuroectoderm progenitor cells can be obtained within 4 to 6 days by using E/NCad-Fc based monolayer-forming ESCs and iPSCs culture protocol and standard neurogenic cocktail treatment, which is a great advantage for quick generation of the target cells for application in regenerative medicine.

Some cells release 90 kDa fragment of soluble NCad (sNCad), and NCad-Fc was used to mimic sNCad response on neurite development [118]. Application of NCad-Fc by Doherty *et al.* with cerebral neurons showed that NCad-Fc initiated neurite outgrowth in a FGF receptor depended manner [111,119] suggesting that NCad-Fc can be utilized for controlling FGF receptor signaling pathway to facilitate relevant neuronal development events. Using mouse E12.5 ventral spinal cord explants as a convenient model Marthiens *et al.* showed that the axons formed contacts along the axon-shaft by long filopodia-like processes on NCad-Fc matrix [120]. They further showed that growth cones preferentially interact with cad-11 or NCad-Fc when progressing on this substratum whereas it differs on laminin. This study proved direct involvement of cadherin-11 and NCad in peripheral nervous system establishment from embryonic tissues [120].

Not only for neuronal population related regenerative medicine, NCad-Fc also showed potentials for application in other tissues as well, for example myogenesis related issues. Charrasse *et al.* used NCad-Fc to mimic NCad binding effect for myogenic differentiation [121]. They showed that NCad-Fc based NCad–dependent cell–cell adhesion triggers RhoA GTPase activity, which is essential for myogenic differentiation. Activity and expression of SRF, a transcription factor that binds to the promoter regions of muscle-specific genes [122,123] and controls the expression of MyoD, is controlled by RhoA. In turn, MyoD binds to the promoter region of skeletal muscle activating genes in mesenchymal cells and convert them to skeletal myoblasts [124,125,126]. These findings demonstrated that N-cadherin–dependent adhesion event that regulates the RhoA/SRF pathway to trigger myogenesis can be harnessed by NCad-Fc matrix and therefore such technology is holding great promises for using in relevant regenerative medicine protocols.

10. Conclusion

To design an efficient biomaterial capable of maintaining and stewarding specific cell phenotypes critical for the development, homeostasis, differentiation, and regeneration of tissues, the material must have a high degree of selective recognition property to the desired cells. As well, such a biomaterial should be devoid of unexpected stimulation characteristics to the cells that can be hazardous to them or to the desired results of the protocols. Being the intrinsic component of cellular milieu, proteins are highly desirable molecules to be used in regenerative medicine and tissue engineering technology. Their 3D conformation made them perfectly fit in the cell-biology and ensuring that only specific function to the experimental cells has been achieved. The natural homeostasis properties of cells can adequately remove these proteins once they are used up without exerting any unnatural effect or stress to the cells. Expressing and purifying large protein with proper 3D conformation is extremely challenging therefore mimetic peptide technology has been becoming popular. These small peptide sequences represent small functional domain of the relevant proteins, albeit not with the native 3D structure of the parent protein molecule. While most cases they are being generated using artificial synthetic technology in test tubes, however, their purity, reproducibility and yield are major concerns for their confident application in stem cell technology. Additional limitations for mimetic peptides are (1) the restricted size of desired peptides, and (2) inability to provide native post-translational modifications, most of which are critical for proper biofunctionality of the relevant molecule. Therefore mimetic peptides cannot and do not behave identically as their natural parent protein. On the contrary, Fc-chimeric proteins can be generated with high degree of reproducibility with identical molecular properties using the natural cellular readout process from the DNA template. The additional stability of the target protein instigated by the presence of Fc domain is significantly advantageous for higher yield of the tailored chimeric protein. The intrinsic property of Fc domain to form homodimer is beneficial to keep the target chimeric protein in soluble form. On the other hand, the natural affinity of Fc domain to bind with Protein A or Protein G is a technical boon for convenient purification of the target protein without fusion of any secondary bait to the amino acid sequence, which often create complex situation for getting rid of them at the later stage of the processing to harvest only the desired designed protein. Directional binding of Fc domain with the polystyrene or hydrophobic surface and catering the functional protein outwards is also an intrinsic benefit for using this class of chimeric proteins for obtaining higher functional efficacy of the applied biomaterials. Since the specific homophilic interactions between cadherins mediate cell attachment therefore specific cadherin isoform-expressing cells can be purified by using the relevant cadherin-Fc biomaterial as surface-coating materials. For example, iPS cells express high level of ECad and neuronal cells express NCad therefore, by employing these matrices in different time points of differentiation protocol, the target cells can be purified *in situ* without the necessity of any harsh enzymatic or mechanical treatments. Some of these chimeric proteins are commercially available for application and some are in pipeline, which can be obtained from our laboratory under proper regulatory affairs. Collectively, Fc-chimeric protein-based biomaterials provide distinct advantages for overcoming many existing challenges in stem cell technology and significantly advancing the regenerative medicine and tissue engineering field towards practical application.

Author details

Kakon Nag, Nihad Adnan, Koichi Kutsuzawa and Toshihiro Akaike*

*Address all correspondence to: takaike@bio.titech.ac.jp

Graduate School of Bioscience and Biotechnology, Tokyo Institute of Technology, Nagatsuta-cho, Midori-ku, Yokohama, Japan

References

[1] Alvarez CV, Garcia-Lavandeira M, Garcia-Rendueles ME, Diaz-Rodriguez E, Garcia-Rendueles AR, Perez-Romero S, et al. Defining stem cell types: understanding the therapeutic potential of ESCs, ASCs, and iPS cells. Journal of molecular endocrinology. 2012;49:R89-111.

[2] Xu C, Jiang J, Sottile V, McWhir J, Lebkowski J, Carpenter MK. Immortalized fibroblast-like cells derived from human embryonic stem cells support undifferentiated cell growth. Stem cells. 2004;22:972-80.

[3] D'Amour KA, Agulnick AD, Eliazer S, Kelly OG, Kroon E, Baetge EE. Efficient differentiation of human embryonic stem cells to definitive endoderm. Nature biotechnology. 2005;23:1534-41.

[4] Yao S, Chen S, Clark J, Hao E, Beattie GM, Hayek A, et al. Long-term self-renewal and directed differentiation of human embryonic stem cells in chemically defined conditions. Proceedings of the National Academy of Sciences of the United States of America. 2006;103:6907-12.

[5] Grigoriadis AE, Kennedy M, Bozec A, Brunton F, Stenbeck G, Park IH, et al. Directed differentiation of hematopoietic precursors and functional osteoclasts from human ES and iPS cells. Blood. 2010;115:2769-76.

[6] Lee G, Chambers SM, Tomishima MJ, Studer L. Derivation of neural crest cells from human pluripotent stem cells. Nature protocols. 2010;5:688-701.

[7] Szabo E, Rampalli S, Risueno RM, Schnerch A, Mitchell R, Fiebig-Comyn A, et al. Direct conversion of human fibroblasts to multilineage blood progenitors. Nature. 2010;468:521-6.

[8] Dar A, Domev H, Ben-Yosef O, Tzukerman M, Zeevi-Levin N, Novak A, et al. Multipotent vasculogenic pericytes from human pluripotent stem cells promote recovery of murine ischemic limb. Circulation. 2012;125:87-99.

[9] Thomson JA, Itskovitz-Eldor J, Shapiro SS, Waknitz MA, Swiergiel JJ, Marshall VS, et al. Embryonic stem cell lines derived from human blastocysts. Science. 1998;282:1145-7.

[10] Takahashi K, Yamanaka S. Induction of pluripotent stem cells from mouse embryonic and adult fibroblast cultures by defined factors. Cell. 2006;126:663-76.

[11] Wang L, Li L, Menendez P, Cerdan C, Bhatia M. Human embryonic stem cells maintained in the absence of mouse embryonic fibroblasts or conditioned media are capable of hematopoietic development. Blood. 2005;105:4598-603.

[12] Levenstein ME, Ludwig TE, Xu RH, Llanas RA, VanDenHeuvel-Kramer K, Manning D, et al. Basic fibroblast growth factor support of human embryonic stem cell self-renewal. Stem cells. 2006;24:568-74.

[13] Ludwig TE, Bergendahl V, Levenstein ME, Yu J, Probasco MD, Thomson JA. Feeder-independent culture of human embryonic stem cells. Nature methods. 2006;3:637-46.

[14] Xu C, Inokuma MS, Denham J, Golds K, Kundu P, Gold JD, et al. Feeder-free growth of undifferentiated human embryonic stem cells. Nature biotechnology. 2001;19:971-4.

[15] Stewart MH, Bendall SC, Bhatia M. Deconstructing human embryonic stem cell cultures: niche regulation of self-renewal and pluripotency. Journal of molecular medicine. 2008;86:875-86.

[16] Kleinman HK, McGarvey ML, Liotta LA, Robey PG, Tryggvason K, Martin GR. Isolation and characterization of type IV procollagen, laminin, and heparan sulfate proteoglycan from the EHS sarcoma. Biochemistry. 1982;21:6188-93.

[17] Kleinman HK, McGarvey ML, Hassell JR, Star VL, Cannon FB, Laurie GW, et al. Basement membrane complexes with biological activity. Biochemistry. 1986;25:312-8.

[18] Vukicevic S, Kleinman HK, Luyten FP, Roberts AB, Roche NS, Reddi AH. Identification of multiple active growth factors in basement membrane Matrigel suggests caution in interpretation of cellular activity related to extracellular matrix components. Experimental cell research. 1992;202:1-8.

[19] Kleinman HK, Martin GR. Matrigel: basement membrane matrix with biological activity. Seminars in cancer biology. 2005;15:378-86.

[20] Carlson J, Garg R, Compton SR, Zeiss C, Uchio E. Poliomyelitis in SCID Mice Following Injection of Basement Membrane Matrix Contaminated with Lactate Dehydrogenase-elevating Virus. J Am Assoc Lab Anim. 2008;47:80-1.

[21] Hayashi Y, Furue MK, Okamoto T, Ohnuma K, Myoishi Y, Fukuhara Y, et al. Integrins regulate mouse embryonic stem cell self-renewal. Stem cells. 2007;25:3005-15.

[22] Domogatskaya A, Rodin S, Boutaud A, Tryggvason K. Laminin-511 but not -332, -111, or -411 enables mouse embryonic stem cell self-renewal in vitro. Stem cells. 2008;26:2800-9.

[23] Ido H, Ito S, Taniguchi Y, Hayashi M, Sato-Nishiuchi R, Sanzen N, et al. Laminin isoforms containing the gamma3 chain are unable to bind to integrins due to the absence of the glutamic acid residue conserved in the C-terminal regions of the gamma1 and gamma2 chains. The Journal of biological chemistry. 2008;283:28149-57.

[24] Miyazaki T, Futaki S, Hasegawa K, Kawasaki M, Sanzen N, Hayashi M, et al. Recombinant human laminin isoforms can support the undifferentiated growth of human embryonic stem cells. Biochemical and biophysical research communications. 2008;375:27-32.

[25] Nagaoka M, Jiang HL, Hoshiba T, Akaike T, Cho CS. Application of recombinant fusion proteins for tissue engineering. Annals of biomedical engineering. 2010;38:683-93.

[26] Collier JH, Segura T. Evolving the use of peptides as components of biomaterials. Biomaterials. 2011;32:4198-204.

[27] Kim SH, Turnbull J, Guimond S. Extracellular matrix and cell signalling: the dynamic cooperation of integrin, proteoglycan and growth factor receptor. The Journal of endocrinology. 2011;209:139-51.

[28] Assenberg R, Wan PT, Geisse S, Mayr LM. Advances in recombinant protein expression for use in pharmaceutical research. Current opinion in structural biology. 2013;23:393-402.

[29] Yue XS, Murakami Y, Tamai T, Nagaoka M, Cho CS, Ito Y, et al. A fusion protein N-cadherin-Fc as an artificial extracellular matrix surface for maintenance of stem cell features. Biomaterials. 2010;31:5287-96.

[30] Haque A, Hexig B, Meng Q, Hossain S, Nagaoka M, Akaike T. The effect of recombinant E-cadherin substratum on the differentiation of endoderm-derived hepatocyte-like cells from embryonic stem cells. Biomaterials. 2011;32:2032-42.

[31] Yu M, Du F, Ise H, Zhao W, Zhang Y, Yu Y, et al. Preparation and characterization of a VEGF-Fc fusion protein matrix for enhancing HUVEC growth. Biotechnol Lett. 2012;34:1765-71.

[32] Nelson AL, Reichert JM. Development trends for therapeutic antibody fragments. Nature biotechnology. 2009;27:331-7.

[33] Zhang J, Carter J, Siu S, O'Neill JW, Gates AH, Delaney J, et al. Fusion partners as a tool for the expression of difficult proteins in mammalian cells. Current pharmaceutical biotechnology. 2010;11:241-5.

[34] Beck A, Reichert JM. Therapeutic Fc-fusion proteins and peptides as successful alternatives to antibodies. mAbs. 2011;3:415-6.

[35] Xiao K, Allison DF, Buckley KM, Kottke MD, Vincent PA, Faundez V, et al. Cellular levels of p120 catenin function as a set point for cadherin expression levels in microvascular endothelial cells. The Journal of cell biology. 2003;163:535-45.

[36] Ireton RC, Davis MA, van Hengel J, Mariner DJ, Barnes K, Thoreson MA, et al. A novel role for p120 catenin in E-cadherin function. The Journal of cell biology. 2002;159:465-76.

[37] Thoreson MA, Anastasiadis PZ, Daniel JM, Ireton RC, Wheelock MJ, Johnson KR, et al. Selective uncoupling of p120(ctn) from E-cadherin disrupts strong adhesion. The Journal of cell biology. 2000;148:189-202.

[38] Cavey M, Rauzi M, Lenne PF, Lecuit T. A two-tiered mechanism for stabilization and immobilization of E-cadherin. Nature. 2008;453:751-6.

[39] Yamada S, Pokutta S, Drees F, Weis WI, Nelson WJ. Deconstructing the cadherin-catenin-actin complex. Cell. 2005;123:889-901.

[40] Palacios F, Tushir JS, Fujita Y, D'Souza-Schorey C. Lysosomal targeting of E-cadherin: a unique mechanism for the down-regulation of cell-cell adhesion during epithelial to mesenchymal transitions. Molecular and cellular biology. 2005;25:389-402.

[41] Balzac F, Avolio M, Degani S, Kaverina I, Torti M, Silengo L, et al. E-cadherin endocytosis regulates the activity of Rap1: a traffic light GTPase at the crossroads between cadherin and integrin function. Journal of cell science. 2005;118:4765-83.

[42] Kobielak A, Pasolli HA, Fuchs E. Mammalian formin-1 participates in adherens junctions and polymerization of linear actin cables. Nature cell biology. 2004;6:21-30.

[43] Xu J, Zhu C, Zhang Y, Jiang N, Li S, Su Z, et al. hE-cadherin-Fc fusion protein coated surface enhances the adhesion and proliferation of human mesenchymal stem cells. Colloids and surfaces B, Biointerfaces. 2013;109:97-102.

[44] Li L, Bennett SA, Wang L. Role of E-cadherin and other cell adhesion molecules in survival and differentiation of human pluripotent stem cells. Cell adhesion & migration. 2012;6:59-70.

[45] Redmer T, Diecke S, Grigoryan T, Quiroga-Negreira A, Birchmeier W, Besser D. E-cadherin is crucial for embryonic stem cell pluripotency and can replace OCT4 during somatic cell reprogramming. EMBO reports. 2011;12:720-6.

[46] Oda H, Takeichi M. Evolution: structural and functional diversity of cadherin at the adherens junction. The Journal of cell biology. 2011;193:1137-46.

[47] Niessen CM, Leckband D, Yap AS. Tissue organization by cadherin adhesion molecules: dynamic molecular and cellular mechanisms of morphogenetic regulation. Physiological reviews. 2011;91:691-731.

[48] Riethmacher D, Brinkmann V, Birchmeier C. A targeted mutation in the mouse E-cadherin gene results in defective preimplantation development. Proceedings of the National Academy of Sciences of the United States of America. 1995;92:855-9.

[49] Larue L, Ohsugi M, Hirchenhain J, Kemler R. E-cadherin null mutant embryos fail to form a trophectoderm epithelium. Proceedings of the National Academy of Sciences of the United States of America. 1994;91:8263-7.

[50] Nagaoka M, Ise H, Akaike T. Immobilized E-cadherin model can enhance cell attachment and differentiation of primary hepatocytes but not proliferation. Biotechnol Lett. 2002;24:1857-62.

[51] Turkova J. Oriented immobilization of biologically active proteins as a tool for revealing protein interactions and function. Journal of chromatography B, Biomedical sciences and applications. 1999;722:11-31.

[52] Awata R, Sawai H, Imai K, Terada K, Senoo H, Sugiyama T. Morphological comparison and functional reconstitution of rat hepatic parenchymal cells on various matrices. Journal of gastroenterology and hepatology. 1998;13 Suppl:S55-61.

[53] Zhang X, Wharton W, Donovan M, Coppola D, Croxton R, Cress WD, et al. Density-dependent growth inhibition of fibroblasts ectopically expressing p27(kip1). Molecular biology of the cell. 2000;11:2117-30.

[54] Kojima N, Kinoshita T, Kamiya A, Nakamura K, Nakashima K, Taga T, et al. Cell density-dependent regulation of hepatic development by a gp130-independent pathway. Biochemical and biophysical research communications. 2000;277:152-8.

[55] Nagaoka M, Koshimizu U, Yuasa S, Hattori F, Chen H, Tanaka T, et al. E-cadherin-coated plates maintain pluripotent ES cells without colony formation. PloS one. 2006;1:e15.

[56] Nagaoka M, Si-Tayeb K, Akaike T, Duncan SA. Culture of human pluripotent stem cells using completely defined conditions on a recombinant E-cadherin substratum. BMC developmental biology. 2010;10:60.

[57] Nagaoka M, Hagiwara Y, Takemura K, Murakami Y, Li J, Duncan SA, et al. Design of the artificial acellular feeder layer for the efficient propagation of mouse embryonic stem cells. The Journal of biological chemistry. 2008;283:26468-76.

[58] Horie M, Ito A, Kiyohara T, Kawabe Y, Kamihira M. E-cadherin gene-engineered feeder systems for supporting undifferentiated growth of mouse embryonic stem cells. Journal of bioscience and bioengineering. 2010;110:582-7.

[59] Ludwig TE, Levenstein ME, Jones JM, Berggren WT, Mitchen ER, Frane JL, et al. Derivation of human embryonic stem cells in defined conditions. Nature biotechnology. 2006;24:185-7.

[60] Nagaoka M, Ise H, Harada I, Koshimizu U, Maruyama A, Akaike T. Embryonic undifferentiated cells show scattering activity on a surface coated with immobilized E-cadherin. Journal of cellular biochemistry. 2008;103:296-310.

[61] Larue L, Antos C, Butz S, Huber O, Delmas V, Dominis M, et al. A role for cadherins in tissue formation. Development. 1996;122:3185-94.

[62] Hyafil F, Babinet C, Jacob F. Cell-cell interactions in early embryogenesis: a molecular approach to the role of calcium. Cell. 1981;26:447-54.

[63] Harb N, Archer TK, Sato N. The Rho-Rock-Myosin signaling axis determines cell-cell integrity of self-renewing pluripotent stem cells. PloS one. 2008;3:e3001.

[64] Kelly KF, Ng DY, Jayakumaran G, Wood GA, Koide H, Doble BW. beta-catenin enhances Oct-4 activity and reinforces pluripotency through a TCF-independent mechanism. Cell stem cell. 2011;8:214-27.

[65] Haque A, Yue XS, Motazedian A, Tagawa Y, Akaike T. Characterization and neural differentiation of mouse embryonic and induced pluripotent stem cells on cadherin-based substrata. Biomaterials. 2012;33:5094-106.

[66] Zhang Z, Gao Y, Gordon A, Wang ZZ, Qian Z, Wu WS. Efficient generation of fully reprogrammed human iPS cells via polycistronic retroviral vector and a new cocktail of chemical compounds. PloS one. 2011;6:e26592.

[67] Liao B, Bao X, Liu L, Feng S, Zovoilis A, Liu W, et al. MicroRNA cluster 302-367 enhances somatic cell reprogramming by accelerating a mesenchymal-to-epithelial transition. The Journal of biological chemistry. 2011;286:17359-64.

[68] Kim H, Lee G, Ganat Y, Papapetrou EP, Lipchina I, Socci ND, et al. miR-371-3 expression predicts neural differentiation propensity in human pluripotent stem cells. Cell stem cell. 2011;8:695-706.

[69] Xu N, Papagiannakopoulos T, Pan G, Thomson JA, Kosik KS. MicroRNA-145 regulates OCT4, SOX2, and KLF4 and represses pluripotency in human embryonic stem cells. Cell. 2009;137:647-58.

[70] Park IH, Zhao R, West JA, Yabuuchi A, Huo H, Ince TA, et al. Reprogramming of human somatic cells to pluripotency with defined factors. Nature. 2008;451:141-6.

[71] Aasen T, Raya A, Barrero MJ, Garreta E, Consiglio A, Gonzalez F, et al. Efficient and rapid generation of induced pluripotent stem cells from human keratinocytes. Nature biotechnology. 2008;26:1276-84.

[72] Yu J, Vodyanik MA, Smuga-Otto K, Antosiewicz-Bourget J, Frane JL, Tian S, et al. Induced pluripotent stem cell lines derived from human somatic cells. Science. 2007;318:1917-20.

[73] Takahashi K, Tanabe K, Ohnuki M, Narita M, Ichisaka T, Tomoda K, et al. Induction of pluripotent stem cells from adult human fibroblasts by defined factors. Cell. 2007;131:861-72.

[74] Li R, Liang J, Ni S, Zhou T, Qing X, Li H, et al. A mesenchymal-to-epithelial transition initiates and is required for the nuclear reprogramming of mouse fibroblasts. Cell stem cell. 2010;7:51-63.

[75] Eastham AM, Spencer H, Soncin F, Ritson S, Merry CL, Stern PL, et al. Epithelial-mesenchymal transition events during human embryonic stem cell differentiation. Cancer research. 2007;67:11254-62.

[76] Hogan C, Serpente N, Cogram P, Hosking CR, Bialucha CU, Feller SM, et al. Rap1 regulates the formation of E-cadherin-based cell-cell contacts. Molecular and cellular biology. 2004;24:6690-700.

[77] Chen T, Yuan D, Wei B, Jiang J, Kang J, Ling K, et al. E-cadherin-mediated cell-cell contact is critical for induced pluripotent stem cell generation. Stem cells. 2010;28:1315-25.

[78] Loh YH, Wu Q, Chew JL, Vega VB, Zhang W, Chen X, et al. The Oct4 and Nanog transcription network regulates pluripotency in mouse embryonic stem cells. Nature genetics. 2006;38:431-40.

[79] Singh A, Suri S, Lee T, Chilton JM, Cooke MT, Chen W, et al. Adhesion strength-based, label-free isolation of human pluripotent stem cells. Nature methods. 2013;10:438-44.

[80] Villa-Diaz LG, Ross AM, Lahann J, Krebsbach PH. Concise review: The evolution of human pluripotent stem cell culture: from feeder cells to synthetic coatings. Stem cells. 2013;31:1-7.

[81] Sundberg M, Jansson L, Ketolainen J, Pihlajamaki H, Suuronen R, Skottman H, et al. CD marker expression profiles of human embryonic stem cells and their neural derivatives, determined using flow-cytometric analysis, reveal a novel CD marker for exclusion of pluripotent stem cells. Stem cell research. 2009;2:113-24.

[82] Pruszak J, Sonntag KC, Aung MH, Sanchez-Pernaute R, Isacson O. Markers and methods for cell sorting of human embryonic stem cell-derived neural cell populations. Stem cells. 2007;25:2257-68.

[83] Reid RA, Hemperly JJ. Human N-cadherin: nucleotide and deduced amino acid sequence. Nucleic acids research. 1990;18:5896.

[84] Walsh FS, Barton CH, Putt W, Moore SE, Kelsell D, Spurr N, et al. N-cadherin gene maps to human chromosome 18 and is not linked to the E-cadherin gene. Journal of neurochemistry. 1990;55:805-12.

[85] Tamura K, Shan WS, Hendrickson WA, Colman DR, Shapiro L. Structure-function analysis of cell adhesion by neural (N-) cadherin. Neuron. 1998;20:1153-63.

[86] Shan WS, Tanaka H, Phillips GR, Arndt K, Yoshida M, Colman DR, et al. Functional cis-heterodimers of N- and R-cadherins. The Journal of cell biology. 2000;148:579-90.

[87] Overduin M, Harvey TS, Bagby S, Tong KI, Yau P, Takeichi M, et al. Solution structure of the epithelial cadherin domain responsible for selective cell adhesion. Science. 1995;267:386-9.

[88] Hatta K, Takeichi M. Expression of N-cadherin adhesion molecules associated with early morphogenetic events in chick development. Nature. 1986;320:447-9.

[89] Warga RM, Kane DA. A role for N-cadherin in mesodermal morphogenesis during gastrulation. Developmental biology. 2007;310:211-25.

[90] Radice GL, Rayburn H, Matsunami H, Knudsen KA, Takeichi M, Hynes RO. Developmental defects in mouse embryos lacking N-cadherin. Developmental biology. 1997;181:64-78.

[91] Takeichi M. Morphogenetic roles of classic cadherins. Current opinion in cell biology. 1995;7:619-27.

[92] Xu L, Overbeek PA, Reneker LW. Systematic analysis of E-, N- and P-cadherin expression in mouse eye development. Experimental eye research. 2002;74:753-60.

[93] Linask KK. N-cadherin localization in early heart development and polar expression of Na+,K(+)-ATPase, and integrin during pericardial coelom formation and epithelialization of the differentiating myocardium. Developmental biology. 1992;151:213-24.

[94] George-Weinstein M, Gerhart J, Blitz J, Simak E, Knudsen KA. N-cadherin promotes the commitment and differentiation of skeletal muscle precursor cells. Developmental biology. 1997;185:14-24.

[95] Takeichi M. The cadherins: cell-cell adhesion molecules controlling animal morphogenesis. Development. 1988;102:639-55.

[96] Hatta K, Takagi S, Fujisawa H, Takeichi M. Spatial and temporal expression pattern of N-cadherin cell adhesion molecules correlated with morphogenetic processes of chicken embryos. Developmental biology. 1987;120:215-27.

[97] Puch S, Armeanu S, Kibler C, Johnson KR, Muller CA, Wheelock MJ, et al. N-cadherin is developmentally regulated and functionally involved in early hematopoietic cell differentiation. Journal of cell science. 2001;114:1567-77.

[98] Olson EN. Interplay between proliferation and differentiation within the myogenic lineage. Developmental biology. 1992;154:261-72.

[99] Ludolph DC, Konieczny SF. Transcription factor families: muscling in on the myogenic program. FASEB journal : official publication of the Federation of American Societies for Experimental Biology. 1995;9:1595-604.

[100] Knudsen KA, Myers L, McElwee SA. A role for the Ca2(+)-dependent adhesion molecule, N-cadherin, in myoblast interaction during myogenesis. Experimental cell research. 1990;188:175-84.

[101] Mege RM, Goudou D, Diaz C, Nicolet M, Garcia L, Geraud G, et al. N-cadherin and N-CAM in myoblast fusion: compared localisation and effect of blockade by peptides and antibodies. Journal of cell science. 1992;103 (Pt 4):897-906.

[102] Soler AP, Knudsen KA. N-cadherin involvement in cardiac myocyte interaction and myofibrillogenesis. Developmental biology. 1994;162:9-17.

[103] Brand-Saberi B, Gamel AJ, Krenn V, Muller TS, Wilting J, Christ B. N-cadherin is involved in myoblast migration and muscle differentiation in the avian limb bud. Developmental biology. 1996;178:160-73.

[104] Panda DK, Miao D, Lefebvre V, Hendy GN, Goltzman D. The transcription factor SOX9 regulates cell cycle and differentiation genes in chondrocytic CFK2 cells. The Journal of biological chemistry. 2001;276:41229-36.

[105] Tuan RS. Cellular signaling in developmental chondrogenesis: N-cadherin, Wnts, and BMP-2. The Journal of bone and joint surgery American volume. 2003;85-A Suppl 2:137-41.

[106] Tufan AC, Tuan RS. Wnt regulation of limb mesenchymal chondrogenesis is accompanied by altered N-cadherin-related functions. FASEB journal : official publication of the Federation of American Societies for Experimental Biology. 2001;15:1436-8.

[107] Ferrari SL, Traianedes K, Thorne M, Lafage-Proust MH, Genever P, Cecchini MG, et al. A role for N-cadherin in the development of the differentiated osteoblastic phenotype. Journal of bone and mineral research : the official journal of the American Society for Bone and Mineral Research. 2000;15:198-208.

[108] Marie PJ. Role of N-cadherin in bone formation. Journal of cellular physiology. 2002;190:297-305.

[109] Nieto MA. The early steps of neural crest development. Mechanisms of development. 2001;105:27-35.

[110] Pla P, Moore R, Morali OG, Grille S, Martinozzi S, Delmas V, et al. Cadherins in neural crest cell development and transformation. Journal of cellular physiology. 2001;189:121-32.

[111] Doherty P, Walsh FS. CAM-FGF Receptor Interactions: A Model for Axonal Growth. Molecular and cellular neurosciences. 1996;8:99-111.

[112] Williams EJ, Williams G, Howell FV, Skaper SD, Walsh FS, Doherty P. Identification of an N-cadherin motif that can interact with the fibroblast growth factor receptor

and is required for axonal growth. The Journal of biological chemistry. 2001;276:43879-86.

[113] Wheelock MJ, Johnson KR. Cadherin-mediated cellular signaling. Current opinion in cell biology. 2003;15:509-14.

[114] Kim JB, Islam S, Kim YJ, Prudoff RS, Sass KM, Wheelock MJ, et al. N-Cadherin extracellular repeat 4 mediates epithelial to mesenchymal transition and increased motility. The Journal of cell biology. 2000;151:1193-206.

[115] Kashima T, Nakamura K, Kawaguchi J, Takanashi M, Ishida T, Aburatani H, et al. Overexpression of cadherins suppresses pulmonary metastasis of osteosarcoma in vivo. International journal of cancer Journal international du cancer. 2003;104:147-54.

[116] Lambert M, Padilla F, Mege RM. Immobilized dimers of N-cadherin-Fc chimera mimic cadherin-mediated cell contact formation: contribution of both outside-in and inside-out signals. Journal of cell science. 2000;113 (Pt 12):2207-19.

[117] Nagaoka M, Akaike T. Single amino acid substitution in the mouse IgG1 Fc region induces drastic enhancement of the affinity to protein A. Protein engineering. 2003;16:243-5.

[118] Derycke L, Morbidelli L, Ziche M, De Wever O, Bracke M, Van Aken E. Soluble N-cadherin fragment promotes angiogenesis. Clinical & experimental metastasis. 2006;23:187-201.

[119] Utton MA, Eickholt B, Howell FV, Wallis J, Doherty P. Soluble N-cadherin stimulates fibroblast growth factor receptor dependent neurite outgrowth and N-cadherin and the fibroblast growth factor receptor co-cluster in cells. Journal of neurochemistry. 2001;76:1421-30.

[120] Marthiens V, Gavard J, Padilla F, Monnet C, Castellani V, Lambert M, et al. A novel function for cadherin-11 in the regulation of motor axon elongation and fasciculation. Molecular and cellular neurosciences. 2005;28:715-26.

[121] Charrasse S, Meriane M, Comunale F, Blangy A, Gauthier-Rouviere C. N-cadherin-dependent cell-cell contact regulates Rho GTPases and beta-catenin localization in mouse C2C12 myoblasts. The Journal of cell biology. 2002;158:953-65.

[122] Gauthier-Rouviere C, Vandromme M, Tuil D, Lautredou N, Morris M, Soulez M, et al. Expression and activity of serum response factor is required for expression of the muscle-determining factor MyoD in both dividing and differentiating mouse C2C12 myoblasts. Molecular biology of the cell. 1996;7:719-29.

[123] Wei L, Zhou W, Croissant JD, Johansen FE, Prywes R, Balasubramanyam A, et al. RhoA signaling via serum response factor plays an obligatory role in myogenic differentiation. The Journal of biological chemistry. 1998;273:30287-94.

[124] Lassar AB, Paterson BM, Weintraub H. Transfection of a DNA locus that mediates the conversion of 10T1/2 fibroblasts to myoblasts. Cell. 1986;47:649-56.

[125] Tapscott SJ, Davis RL, Thayer MJ, Cheng PF, Weintraub H, Lassar AB. MyoD1: a nuclear phosphoprotein requiring a Myc homology region to convert fibroblasts to myoblasts. Science. 1988;242:405-11.

[126] Choi J, Costa ML, Mermelstein CS, Chagas C, Holtzer S, Holtzer H. MyoD converts primary dermal fibroblasts, chondroblasts, smooth muscle, and retinal pigmented epithelial cells into striated mononucleated myoblasts and multinucleated myotubes. Proceedings of the National Academy of Sciences of the United States of America. 1990;87:7988-92.

Chapter 7

Hepatocyte Selection Medium

Minoru Tomizawa, Fuminobu Shinozaki,
Yasufumi Motoyoshi, Takao Sugiyama,
Shigenori Yamamoto and Makoto Sueishi

Additional information is available at the end of the chapter

http://dx.doi.org/10.5772/58394

1. Introduction

Embryonic stem (ES) cells have the potential to differentiate to hepatocytes [1]. However, the use of ES cells may pose ethical problems because they are derived from human embryos. The use of human induced pluripotent stem (hiPS) cells that have been generated from adult somatic cells [2], on the other hand, does not create ethical controversies. HiPS cells are useful tools in drug discovery and regenerative medicine because they can differentiate into functional somatic cells [3]. If hiPS cells could be differentiated into hepatocytes, they would be useful for transplantation into patients suffering from hepatic failure [4]. Complications such as graft-versus-host disease as well as ethical issues could be avoided because patient-specific somatic cells could be generated from hiPS cells isolated from the patient.

The ES and hiPS cells that survive among the differentiated hepatocytes and are transplanted to patients may be tumorigenic [5]. Therefore, methods need to be developed to eliminate ES and iPS cells from the population of differentiated cells used for transplantation. To overcome these problems, a new medium, called "hepatocyte selection medium" (HSM), has been developed and will be discussed in this chapter [6].

First, pluripotency and tumorigenicity of ES and iPS cells will be discussed [7]. Next, current methods of eliminating pluripotent cells will be outlined [8, 9]. All the cells, including human iPS cells, require glucose and arginine to live [10, 11]. They will die without glucose or arginine. Hepatocytes have enzymes to produce glucose from galactose and arginine from ornithine. The unique features of hepatocytes compared with other cells will next be discussed. It was expected that hepatocytes would survive in a medium without glucose or arginine, and supplemented with galactose and ornithine [12] [13]. After this introduction, the formulation of HSM will be described [14]. Finally, the application of HSM for the selection of cells differentiated from mouse ES and human iPS cells will be presented.

2. Pluripotency and tumorigenicity

The link between pluripotency and tumorigenicity was reported in 1960 based on a study of teratocarcinoma [15]. ES and iPS cells are pluripotent and are capable of self-renewal as well as differentiation into a variety of cell types. Pluripotent cells can, however, be tumorigenic because they proliferate rapidly and exhibit telomerase activity [7]. Therefore, one of the problems faced while using ES and iPS cell-derived cells for transplantation into patients is the risk of tumorigenicity [5]. For example, transplantation of mouse hepatocytes differentiated from ES cells into liver resulted in the formation of teratoma [16]. Tumorigenicity was initially attributed to genomic integration of the viral vectors used for the induction of pluripotency [17]. The Sendai virus was also used to generate iPS cells because it posed no risk of altering the host genome [18]. To reduce this risk, plasmid vectors have been used to introduce reprogramming factors such as Oct3/4, Sox2, Klf4, and c-Myc [19]. In addition, the ES cell-specific microRNA, miR-302, has been used to reduce the iPS cells tumorigenicity by suppressing cyclin E-CDK2 and cyclin D-CDK4/6 [20]. Furthermore, Yakubov et al. introduced RNA synthesized from the cDNA of the four reprogramming transcription factors [21]. Several combinations of reprogramming factors have also been investigated. Nakagawa et al. omitted c-Myc to generate iPS cells, thereby reducing the tumorigenicity because c-Myc is a well-known oncogene [22]. Despite these efforts, the risk of tumorigenicity has not yet been eliminated. The process of pluripotency and tumorigenicity involve self-renewal, proliferation, and active telomerase mechanisms [7]. It is, therefore, necessary to develop methods for the efficient eradication of iPS cells that survive among differentiated somatic cells.

3. Methods of eliminating iPS cells

Flow cytometry, which is commonly used to isolate target cells, was used by Yamamoto et al. to isolate hepatocytes differentiated from the mouse ES cells [8]. These workers generated ES cells expressing green fluorescent protein driven by an albumin promoter/enhancer. However, since albumin is expressed in endodermal cells as well [23], this approach led to isolation of cells other than hepatocytes, such as endodermal cells. Therefore, a different strategy was required to improve hepatocyte isolation. Flow cytometry has also been used to analyze surface antigens specific for hepatocytes. For example, delta-like 1 homolog (DLK1) has been used for isolation of hepatoblasts [9]. The issue with DLK1, however, is that this surface antigen is not expressed in the human adult liver [24]. Therefore, it may not be possible to isolate mature hepatocytes differentiated from hiPS cells using DLK-1 as a marker. In our research we focused on other methods to eliminate iPS or ES cells from heptocytes. Sub-lethal heat shock was shown to induce apoptosis in human ES cells [25], but it might also damage differentiated cells intended for transplantation. Cheng et al. reported the same strategy using suicide genes [26]. They introduced a thymidine kinase gene driven by the Nanog promoter into hiPS cells, which were subsequently ablated by ganciclovir treatment. This method may be ideal for differentiated hepatocytes, which do not express Nanog, but the toxicity of ganciclovir may be a potential issue. Conesa et al. screened a library of 1120 small chemicals to identify molecules that caused mouse ES cells to undergo apoptosis [27], and found that benzethonium chloride and methylbenzethonium induced apoptosis in hiPS and mouse ES cells but not in human fibroblasts or mouse embryonic fibroblasts. Both reagents are quaternary ammonium salts used as antimicrobial agents; they are also used in cancer therapy and may have damaging effects on hepatocytes. *N*-oleoyl serinol (S18), which is a ceramide analogue, eliminated residual pluripotent cells in embryoid bodies [28]. Interestingly, S18 also promoted neural differentiation of embryoid body-derived cells. This strategy is promising because the reagent not only eradicates undifferentiated cells but also promotes their differentiation into the target cell types.

4. Arginine and urea cycle

Among all the amino acids, the deficiency in arginine is the least tolerated by the cells cultured *in vitro* [29]. Arginine is produced through the urea cycle, which is exclusive to hepatocytes. Indeed, an arginine-deficient medium was the first one used for the hepatocyte selection [10]. Tyrosine also is produced by hepatocytes, and H4 II E, a hepatoma cell line adapted to growth in serum-, arginine-, and tyrosine-free medium, has been established [30]. This cell line expresses ornithine transcarbamylase (OTC) involved in the urea cycle, and phenylalanine hydroxylase (PAH), which catalyzes the synthesis of tyrosine in the liver and kidney [31].

The major role of urea synthesis is the excretion of ammonium ions generated in the process of protein degradation. Urea synthesis is a cyclic process as shown in Figure 1. Ornithine plays the key role in urea synthesis, and OTC mediates the formation of L-citrulline from L-ornithine and carbamoylphosphate. Importantly, this process occurs in liver mitochondria (area bounded with green line in Figure 1). The OTC deficiency, linked to X-chromosome is the cause of hyperammonemia type 2 [32]. The elevated ammonium levels lead to infantile death or mental retardation later in life.

Consequently, it could be expected that hepatocytes can be selected from ES cells in a medium deficient in arginine and tyrosine.

5. Glucose and gluconeogenesis

Glucose is an important source of energy for a majority of cells. Glucose deprivation aids in the hepatocyte selection process because hepatocytes are capable of synthesizing glucose [10]. Pyruvate is the final product of glycolysis, which then enters the tricarboxylic acid cycle. It was shown that pyruvate and glucose deficiency led to neural cell death [11]. Galactose enters glycolysis as a substrate for galactokinase, which is expressed in the liver and kidney [33, 34]. Therefore, it is expected that hepatocytes can survive in a medium deprived of glucose or pyruvate but supplemented with galactose [12] [13].

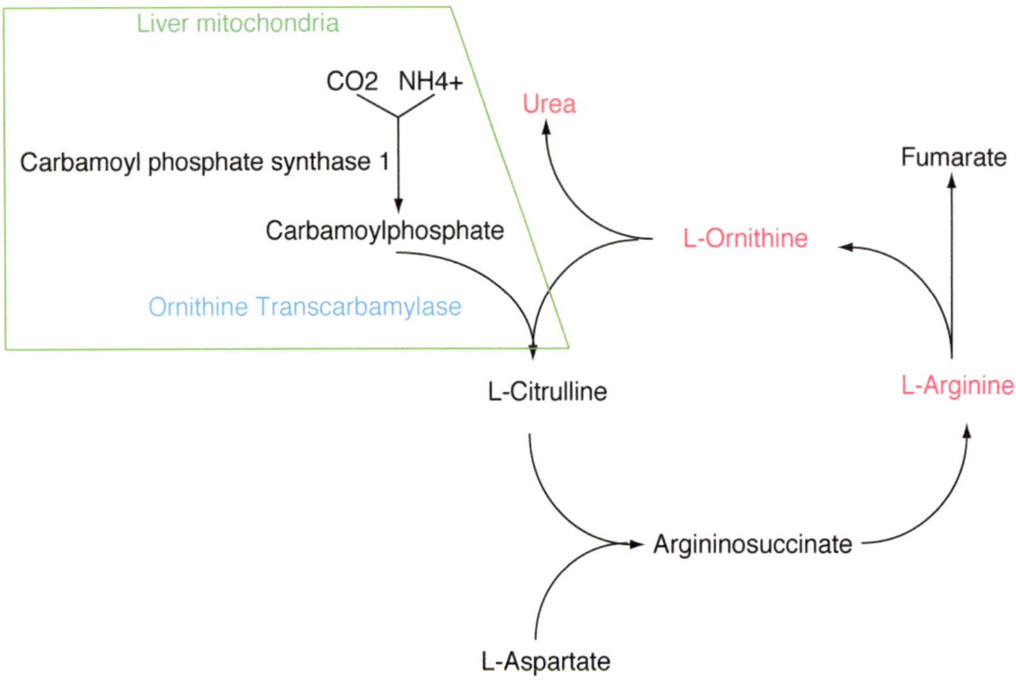

Figure 1. Urea cycle.

Galactose is produced from lactose by hydrolysis in the gastrointestinal tract and is converted to glucose in the liver (Figure 2). Galactokinase catalyzes ATP-dependent phosphorylation of galactose to galactose 1-phosphate which then reacts with uridine diphosphate(UDP)-glucose to produce UDP-galactose converted to UDP-glucose by uridine diposphogalactose 4-epimerase. UDP-glucose is used by glycogen synthase to synthesize glycogen, which is stored in the liver and used as a source of glucose.

Deficiency in the enzymes such as galactokinase, galactose-1-phosphate uridyltransferase, or uridine diphosphogalactose 4-epimerase causes galactosemia. Galactose is then reduced to galactitol, which accumulates in the eye lenses causing cataracts. Deficiency in galactose-1-phosphate uridyltransferase results in accumulation of galactose-1-phosphate and depletion of inorganic phosphate in the liver causing liver failure. This is the reason why children suffering from galactosemia are kept on a galactose-free diet.

6. Hepatocyte selection medium

The hepatocyte selection medium (HSM) was made from powdered amino acids following the formulation of Leibovits-15 medium (Life Technologies, Grand Island, NY). HSM did not contain arginine, tyrosine, glucose, and sodium pyruvate, but was supplemented with galactose (900 mg/L), ornithine (1 mM), glycerol (5 mM), and proline (260 mM) (all from Wako

http://dx.doi.org/10.5772/58394

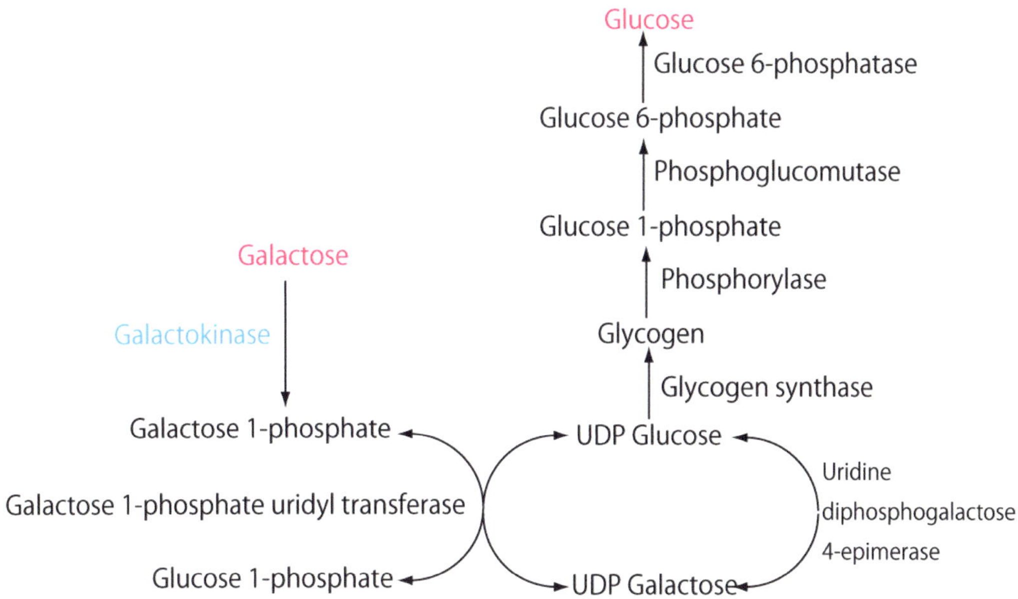

Figure 2. Galactose metabolism.

Pure Chemicals, Osaka, Japan); proline (30 mg/L) was added as a component necessary for DNA synthesis [35]. Aspartic acid as a nonessential amino acid was not included because it can be synthesized from ornithine and arginine. Fetal calf serum (FCS, Life Technologies) at a final concentration of 10% was used to culture mouse ES cells. For human iPS cells, 10% knockout serum replacement (KSR) (Life Technologies) was used instead of FCS to establish xeno-free conditions. Depending on the experiment, FCS and KSR were dialyzed against phosphate buffered saline (PBS) to remove amino acids and glucose.

7. Embryoid bodies in HSM

EB5, a mouse ES cell line provided by Dr. H. Niwa (Center for Developmental Biology, Riken, Kobe, Japan) was maintained in the undifferentiated state in gelatin-coated dishes without feeder cells, in Glasgow minimum essential medium (GMEM) (Sigma Aldrich Japan K.K., Tokyo, Japan) supplemented with 10% FCS (Roche Diagnostics K.K., Tokyo, Japan), 1× nonessential amino acids (NEAA), sodium pyruvate (1 mM), leukemia inhibitory factor (LIF) (1000 U/ml) (Invitrogen Japan, Tokyo, Japan), and 2-mercaptoethanol (0.1 mM) (Wako) [36]. Dissociated ES cells were cultured in hanging drops at a density of 1×10^3 cells per 30 μμl of media without LIF (ESM) to form embryoid bodies. After four days in hanging drop culture, the resulting embryoid bodies were plated onto plastic dishes (Iwaki-Asahi Techno Glass, Tokyo, Japan) precoated with gelatin (Sigma Aldrich). Seven days after their formation, the embryoid bodies transferred to HSM appeared slightly smaller than those in ESM. The cells comprising the embryoid bodies in ESM differentiated to various cell types 28 days after the formation of embryoid bodies.

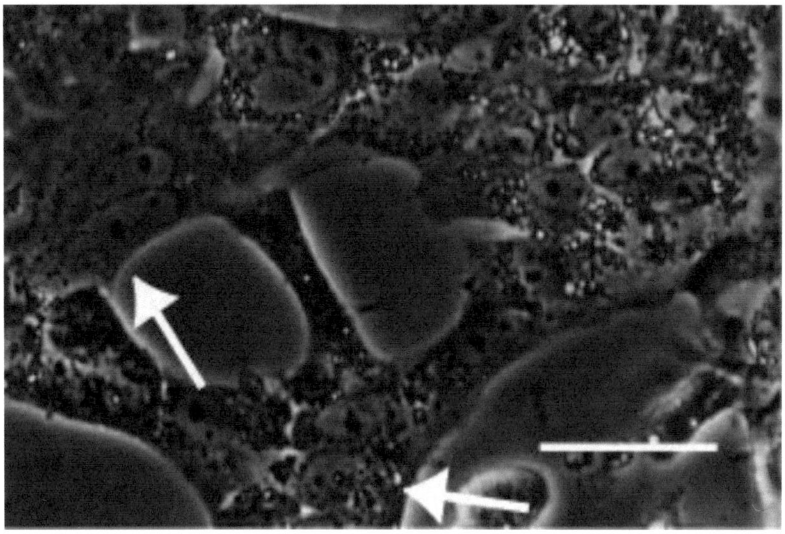

Figure 3. Mouse embryonic stem cells in HSM. Scale bar, 250 μm.

28 days after the formation of embryoid bodies, sizes of colonies in HSM reduced, and the surviving cells appeared cuboidal (Figure 3). Some of these cells were binuclear, which is characteristic of hepatocytes; it was also previously shown that HSM was selective for hepatoblast-like cells [14]. These results suggest that HSM eliminated undifferentiated cells and enriched the population of hepatoblast-like cells.

8. Expression levels of GALK1, GALK2, and OTC

Human fetal and adult hepatocytes express galactokinase and OTC, and would survive in HSM containing galactose and ornithine. If hiPS cells express similar levels of these enzymes, HSM could not be applied for selection of differentiated hepatocytes. Therefore, we compared the expression levels of galactokinase and OTC in hiPS cells with those in human fetal and adult livers. The hiPS cell line 201B7 (RIKEN Cell Bank, Tsukuba, Japan) was cultured feeder-free in ReproFF medium (Reprocell, Yokohama, Japan) in dishes coated with a thin layer of Matrigel (Becton Dickinson, Franklin Lakes, NJ). Two galactokinase isoforms, GALK1 (GenBank: NM_000154) and GALK2 (BC107153), have been identified in humans. The expression levels of GALK1, GALK2, and OTC in the 201B7 cells and fetal and adult livers were compared [6]. The expression levels of these enzymes in the 201B7 cells constituted 22.2% ± 5.0%, 14.2% ± 1.1%, and 1.2% ± 0.2% (mean ± standard deviation) of those in the adult liver, respectively, and the OTC expression was also significantly lower in the 201B7 cells than in the fetal liver. We then cultured 201B7 cells in HSM to assess their survival rates.

9. Human iPS cells in HSM

The 201B7 cells were cultured in 6-well plates coated with Matrigel in the ReproFF medium, which was then changed to HSM (Figure 4). The 201B7 cells started to die and were completely eliminated in three days. Nuclear condensation and fragmentation was observed after staining with hematoxylin and eosin [6]. These nuclei also tested positive by terminal deoxynucleotidyl transferase (TdT)-mediated dUTP nick end labeling (TUNEL). Some of the 201B7 cells that survived in HSM one day after medium change to HSM were immunostained with antibodies against Nanog, SSEA4, and TRA-1-60. The results suggested that the death of undifferentiated 201B7 cells in HSM was caused by apoptosis.

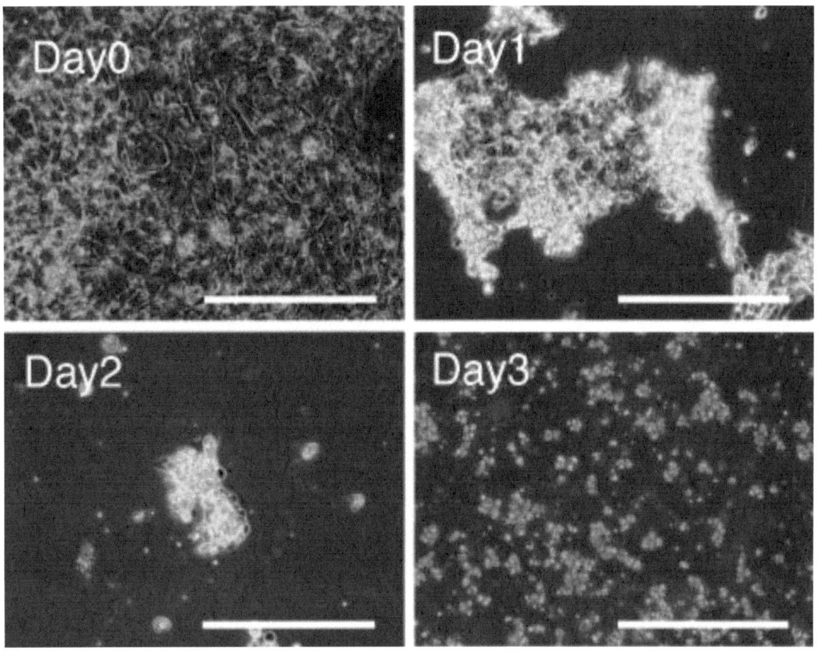

Figure 4. Human iPS cells cultured in HSM. Scale bar, 50 μm. Medium was changed to HSM for human iPS cells in feeder-free culture. All the human iPS cells died on day 3.

10. Primary human hepatocytes

Several protocols for the differentiation of iPS cells to hepatocytes have been reported [3, 37], which describe the differentiation of iPS cells into hepatocyte-like cells which are different from primary human hepatocytes. Recently, a method to generate three-dimensional vascularized liver from iPS cells has been reported [38]. The authors induced hepatic differentiation of human iPS cells by following the protocol described by Si-Tayeb et al [37]. They mixed the iPS cells with vascular endothelial and mesenchymal stem cells, and transplanted them into a mouse brain. This method is sophisticated and promising, but xenograft rejection may be a problem when the generated liver is transplanted to patients with liver failure. Practical methods for the differentiation of human iPS cell to functional hepatocytes are not available. It is therefore necessary to use primary human hepatoctyes as a model of hepatocytes fully differentiated from iPS cells. Hepatocytes were isolated from a fragment of resected donor liver by using 2-step collagenase perfusion [39].

11. Co-culture of human iPS cells and primary human hepatocytes

Methods have not been established regarding hepatocye differentiation from human iPS cells. It is impossible to select hepatocytes differentiated from human iPS cells from the mixture of human iPS cells. Primary human hepatocytes were used as a model of hepatocytes differentiated from human iPS cells. It was expected that human iPS cells and hepatocytes differentiated from them were mixed. Therefore, co-culure of primary human hepatocytes and human iPS cells was used as a model of the mixtures. Primary human hepatocytes were purchased from Lonza (Walkersville, MD) and cultured as per the manufacturer's instructions. Briefly, hepatocytes were thawed and spread at a density of 1.5×10^5 cells/cm^2 onto CellBIND 24-well plates coated with type I collagen from the bovine dermis (Koken Co., Ltd., Tokyo, Japan) and cultured in the hepatocyte culture medium (HCM, Lonza).

The 201B7 cells and human primary hepatocytes were co-cultured as follows: human primary hepatoctyes were cultured in HCM for 24 h as described above. The 201B7 cells were added to the wells at a density of 3×10^4 cells/well. After 24 h of culture in the ReproFF medium, it was changed to HSM. Human primary hepatocytes survived in HSM, while the human 201B7 cells did not (Figure 5).

12. Potential application of HSM

The HSM that we developed can be safely used for the elimination of hiPS cells because it does not contain hazardous reagents or introduce genetic material. Our results show that hiPS cells die after three days of culture in HSM. Prior to performing the experiments, we compared the hiPS cell viability in media containing crude or dialyzed KSR or combination of insulin (10 μM), dexamethasone (10 μM) and aprotitin (5000 U/ml) (IDA) Unexpectedly, the KSR dialysis and IDA had no effect on hiPS cell survival. As expected, primary human hepatocytes survived in HSM as well as in HCM, which is the recommended medium for their culture.

HSM can be used in clinical practices in situations when hepatocytes differentiated from human iPS cells are transplanted to patients suffering from liver failure.

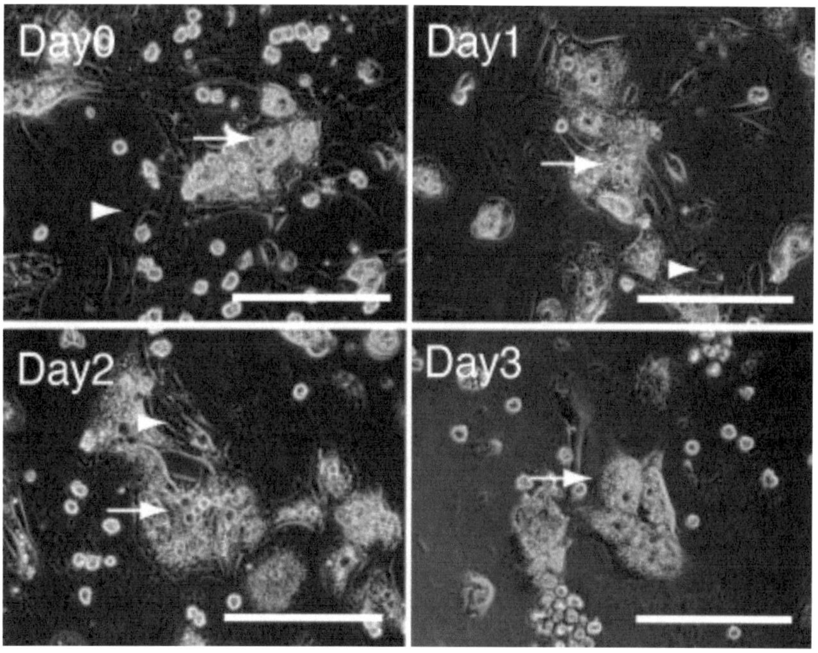

Figure 5. Human iPS cells co-cultured with primary human hepatocytes in HSM. Scale bar, 50 μm; arrow, hepatocytes; arrowhead, 201B7 cells.

13. Conclusion

HSM can be successfully used for the selection of hepatoblast-like cells derived from mouse ES cells. HSM is an ideal medium for the elimination of hiPS cells and the isolation of differentiated hepatocytes without causing any damage. In the future, methods will be established to produce hepatocytes from human iPS cells. Residual human iPS cells are a potential hazard when the hepatocytes will be transplanted for patients with liver insufficiency because the undifferentiated cell harbor tumorigenicity. At that stage, HSM will be an indispensable medium to select hepatocytes differentiated from residual human iPS cells.

Acknowledgements

This work was supported in part by a Research Grant-in-Aid for Scientific Research (C) (Grant No. 23591002) from the Japan Society for the Promotion of Science (JSPS).

Author details

Minoru Tomizawa[1*], Fuminobu Shinozaki[2], Yasufumi Motoyoshi[3], Takao Sugiyama[4], Shigenori Yamamoto[5] and Makoto Sueishi[4]

*Address all correspondence to: nihminor-cib@umin.ac.jp

1 Department of Gastroenterology, National Hospital Organization Shimoshizu Hospital, Yotsukaido City, Japan

2 Department of Radiology, National Hospital Organization Shimoshizu Hospital, Yotsukaido City, Japan

3 Department of Neurology, National Hospital Organization Shimoshizu Hospital, Yotsukaido City, Japan

4 Department of Rheumatology, National Hospital Organization Shimoshizu Hospital, Yotsukaido City, Japan

5 Department of Pediatrics, National Hospital Organization Shimoshizu Hospital, Yotsukaido City, Japan

References

[1] Greenhough S, Bradburn H, Gardner J, Hay DC. Development of an embryoid body-based screening strategy for assessing the hepatocyte differentiation potential of human embryonic stem cells following single-cell dissociation. Cell Reprogram 2013;15(1):9-14.

[2] Takahashi K, Tanabe K, Ohnuki M, Narita M, Ichisaka T, Tomoda K, Yamanaka S. Induction of pluripotent stem cells from adult human fibroblasts by defined factors. Cell 2007;131(5):861-72.

[3] Takayama K, Inamura M, Kawabata K, Sugawara M, Kikuchi K, Higuchi M, Nagamoto Y, Watanabe H, Tashiro K, Sakurai F, Hayakawa T, Furue MK, Mizuguchi H. Generation of metabolically functioning hepatocytes from human pluripotent stem cells by FOXA2 and HNF1alpha transduction. J Hepatol 2012.

[4] am Esch JS, 2nd, Knoefel WT, Klein M, Ghodsizad A, Fuerst G, Poll LW, Piechaczek C, Burchardt ER, Feifel N, Stoldt V, Stockschlader M, Stoecklein N, Tustas RY, Eisenberger CF, Peiper M, Haussinger D, Hosch SB. Portal application of autologous CD133+bone marrow cells to the liver: a novel concept to support hepatic regeneration. Stem Cells 2005;23(4):463-70.

[5] Okita K, Ichisaka T, Yamanaka S. Generation of germline-competent induced pluripotent stem cells. Nature 2007;448(7151):313-7.

[6] Tomizawa M, Shinozaki F, Sugiyama T, Yamamoto S, Sueishi M, Yoshida T. Survival of primary human hepatocytes and death of induced pluripotent stem cells in media lacking glucose and arginine. PLoS One 2013;8(8):e71897.

[7] Kooreman NG, Wu JC. Tumorigenicity of pluripotent stem cells: biological insights from molecular imaging. J R Soc Interface 2010;7 Suppl 6(S753-63.

[8] Yamamoto H, Quinn G, Asari A, Yamanokuchi H, Teratani T, Terada M, Ochiya T. Differentiation of embryonic stem cells into hepatocytes: biological functions and therapeutic application. Hepatology 2003;37(5):983-93.

[9] Tanaka M, Okabe M, Suzuki K, Kamiya Y, Tsukahara Y, Saito S, Miyajima A. Mouse hepatoblasts at distinct developmental stages are characterized by expression of EpCAM and DLK1: drastic change of EpCAM expression during liver development. Mech Dev 2009;126(8-9):665-76.

[10] Leffert HL, Paul D. Studies on primary cultures of differentiated fetal liver cells. J Cell Biol 1972;52(3):559-68.

[11] Matsumoto K, Yamada K, Kohmura E, Kinoshita A, Hayakawa T. Role of pyruvate in ischaemia-like conditions on cultured neurons. Neurol Res 1994;16(6):460-4.

[12] Phillips JW, Jones ME, Berry MN. Implications of the simultaneous occurrence of hepatic glycolysis from glucose and gluconeogenesis from glycerol. Eur J Biochem 2002;269(3):792-7.

[13] Sumida KD, Crandall SC, Chadha PL, Qureshi T. Hepatic gluconeogenic capacity from various precursors in young versus old rats. Metabolism 2002;51(7):876-80.

[14] Tomizawa M, Toyama Y, Ito C, Toshimori K, Iwase K, Takiguchi M, Saisho H, Yokosuka O. Hepatoblast-like cells enriched from mouse embryonic stem cells in medium without glucose, pyruvate, arginine, and tyrosine. Cell Tissue Res 2008;333(1):17-27.

[15] Pierce GB, Jr., Dixon FJ, Jr., Verney EL. Teratocarcinogenic and tissue-forming potentials of the cell types comprising neoplastic embryoid bodies. Lab Invest 1960;9(583-602.

[16] Teramoto K, Hara Y, Kumashiro Y, Chinzei R, Tanaka Y, Shimizu-Saito K, Asahina K, Teraoka H, Arii S. Teratoma formation and hepatocyte differentiation in mouse liver transplanted with mouse embryonic stem cell-derived embryoid bodies. Transplant Proc 2005;37(1):285-6.

[17] Miura K, Okada Y, Aoi T, Okada A, Takahashi K, Okita K, Nakagawa M, Koyanagi M, Tanabe K, Ohnuki M, Ogawa D, Ikeda E, Okano H, Yamanaka S. Variation in the safety of induced pluripotent stem cell lines. Nat Biotechnol 2009;27(8):743-5.

[18] Fusaki N, Ban H, Nishiyama A, Saeki K, Hasegawa M. Efficient induction of transgene-free human pluripotent stem cells using a vector based on Sendai virus, an

RNA virus that does not integrate into the host genome. Proc Jpn Acad Ser B Phys Biol Sci 2009;85(8):348-62.

[19] Okita K, Nakagawa M, Hyenjong H, Ichisaka T, Yamanaka S. Generation of mouse induced pluripotent stem cells without viral vectors. Science 2008;322(5903):949-53.

[20] Lin SL, Ying SY. Mechanism and method for generating tumor-free iPS cells using intronic microRNA miR-302 induction. Methods Mol Biol 2013;936(295-312.

[21] Yakubov E, Rechavi G, Rozenblatt S, Givol D. Reprogramming of human fibroblasts to pluripotent stem cells using mRNA of four transcription factors. Biochem Biophys Res Commun 2010;394(1):189-93.

[22] Nakagawa M, Koyanagi M, Tanabe K, Takahashi K, Ichisaka T, Aoi T, Okita K, Mochiduki Y, Takizawa N, Yamanaka S. Generation of induced pluripotent stem cells without Myc from mouse and human fibroblasts. Nat Biotechnol 2008;26(1):101-6.

[23] Abe K, Niwa H, Iwase K, Takiguchi M, Mori M, Abe SI, Abe K, Yamamura KI. Endoderm-specific gene expression in embryonic stem cells differentiated to embryoid bodies. Exp Cell Res 1996;229(1):27-34.

[24] Yanai H, Nakamura K, Hijioka S, Kamei A, Ikari T, Ishikawa Y, Shinozaki E, Mizunuma N, Hatake K, Miyajima A. Dlk-1, a cell surface antigen on foetal hepatic stem/progenitor cells, is expressed in hepatocellular, colon, pancreas and breast carcinomas at a high frequency. J Biochem 2010;148(1):85-92.

[25] Alekseenko LL, Zemelko VI, Zenin VV, Pugovkina NA, Kozhukharova IV, Kovaleva ZV, Grinchuk TM, Fridlyanskaya, II, Nikolsky NN. Heat shock induces apoptosis in human embryonic stem cells but a premature senescence phenotype in their differentiated progeny. Cell Cycle 2012;11(17):3260-9.

[26] Cheng F, Ke Q, Chen F, Cai B, Gao Y, Ye C, Wang D, Zhang L, Lahn BT, Li W, Xiang AP. Protecting against wayward human induced pluripotent stem cells with a suicide gene. Biomaterials 2012;33(11):3195-204.

[27] Conesa C, Doss MX, Antzelevitch C, Sachinidis A, Sancho J, Carrodeguas JA. Identification of specific pluripotent stem cell death--inducing small molecules by chemical screening. Stem Cell Rev 2012;8(1):116-27.

[28] Bieberich E, Silva J, Wang G, Krishnamurthy K, Condie BG. Selective apoptosis of pluripotent mouse and human stem cells by novel ceramide analogues prevents teratoma formation and enriches for neural precursors in ES cell-derived neural transplants. J Cell Biol 2004;167(4):723-34.

[29] Wheatley DN, Scott L, Lamb J, Smith S. Single amino acid (arginine) restriction: growth and death of cultured HeLa and human diploid fibroblasts. Cell Physiol Biochem 2000;10(1-2):37-55.

[30] Niwa A, Yamamoto K, Sorimachi K, Yasumura Y. Continuous culture of Reuber hepatoma cells in serum-free arginine-, glutamine-and tyrosine-deprived chemically defined medium. In Vitro 1980;16(11):987-93.

[31] McGee MM, Greengard O, Knox WE. The quantitative determination of phenylalanine hydroxylase in rat tissues. Its developmental formation in liver. Biochem J 1972;127(4):669-74.

[32] Kido J, Nakamura K, Mitsubuchi H, Ohura T, Takayanagi M, Matsuo M, Yoshino M, Shigematsu Y, Yorifuji T, Kasahara M, Horikawa R, Endo F. Long-term outcome and intervention of urea cycle disorders in Japan. J Inherit Metab Dis 2012;35(5):777-85.

[33] Ohira RH, Dipple KM, Zhang YH, McCabe ER. Human and murine glycerol kinase: influence of exon 18 alternative splicing on function. Biochem Biophys Res Commun 2005;331(1):239-46.

[34] Ai Y, Jenkins NA, Copeland NG, Gilbert DH, Bergsma DJ, Stambolian D. Mouse galactokinase: isolation, characterization, and location on chromosome 11. Genome Res 1995;5(1):53-9.

[35] Nakamura T, Teramoto H, Tomita Y, Ichihara A. L-proline is an essential amino acid for hepatocyte growth in culture. Biochem Biophys Res Commun 1984;122(3):884-91.

[36] Niwa H, Masui S, Chambers I, Smith AG, Miyazaki J. Phenotypic complementation establishes requirements for specific POU domain and generic transactivation function of Oct-3/4 in embryonic stem cells. Mol Cell Biol 2002;22(5):1526-36.

[37] Si-Tayeb K, Noto FK, Nagaoka M, Li J, Battle MA, Duris C, North PE, Dalton S, Duncan SA. Highly efficient generation of human hepatocyte-like cells from induced pluripotent stem cells. Hepatology 2010;51(1):297-305.

[38] Takebe T, Sekine K, Enomura M, Koike H, Kimura M, Ogaeri T, Zhang RR, Ueno Y, Zheng YW, Koike N, Aoyama S, Adachi Y, Taniguchi H. Vascularized and functional human liver from an iPSC-derived organ bud transplant. Nature 2013;499(7459): 481-4.

[39] Strom SC, Chowdhury JR, Fox IJ. Hepatocyte transplantation for the treatment of human disease. Semin Liver Dis 1999;19(1):39-48.

Section 4

New Stem Cell Models

Chapter 8

Human Pluripotent Stem Cell Applications in Drug Discovery and Toxicology – An overview

Shiva Prasad Potta, Tomo Šarić, Michael Heke,
Harinath Bahudhanapati and Jürgen Hescheler

Additional information is available at the end of the chapter

http://dx.doi.org/10.5772/58485

1. Introduction

Various drugs are being introduced into market for generating beneficial therapeutic effects in humans. The pharmaceutical industry invests about $1.5 billion over the time period of 10-15 years to take a candidate drug from primary screen to market. Unfortunately, many drugs are withdrawn due to side effects associated with off-and on-target toxicity [1]. For example, as many as nine out of ten promising candidates beginning clinical phase I will not achieve marketing approval [2] and only 20% of agents that show efficacy against cardiovascular diseases in preclinical development are licensed after demonstrating sufficient efficacy in phase III testing [3]. The success rate in anticancer drug development process is with 5% of licensed agents even lower. Off-target cardiac toxicity is the most common cause of regulatory delay in approval and market withdrawal of newly developed pharmaceuticals [4, 5]. Drug-induced sudden cardiac death and ventricular arrhythmia caused the withdrawal of more drugs in recent years than any other adverse drug reaction. Moreover, over 100 non-cardiac drugs are suspected to be of high-risk and carry cardiovascular-related black box warnings [6]. Similar considerations are raised concerning arrhythmia and toxicity induced by environmental factors, including industrial chemicals, food additives, cosmetics, and others, as outlined in the European REACH initiative (Registration, Evaluation, Authorization and Restriction of Chemical substances).

Current drug safety evaluation processes that are required for regulatory purposes mostly rely on animal studies and immortalized cell-based assays due to lack of suitable human in vitro cell systems. In Europe, almost 10 million vertebrate animals are used annually for research. Although highly predictive assays involving whole heart or slice preparations and in vivo animal testing remain the standard for preclinical safety pharmacology, this extensive use of animals and their tissues does not eliminate high attrition rates of novel drugs. One of the major reasons for this is limited predictability of existing preclinical animal (and cellular) models for assessment of drug safety and efficacy. Animal models do not always predict the toxicity in humans with sufficient accuracy because of inter-species differences [7]. For example, murine and human hearts greatly differ in some aspects of electrophysiological properties [8]. In addition, inbred animals that are frequently used in these analyses do not mimic the genetic diversity of human population required for accurate prediction of drug responses [9]. Therefore, identification of reliable and robust human cell systems for toxicity assessment has become a driving interest for pharmaceutical industries.

In cardiac area different types of tests are already playing an important role in reducing costs and drug attrition rates. These strategies involve a tiered system which starts with in vitro single cell analyses followed by tests with ex vivo tissues and organs and progresses to in vivo animal models and, finally, clinical trials [10]. The most important in vitro test consists of automated patch-clamp recordings of Chinese hamster ovary (CHO) cells expressing human Ether-à-go-go-Related Gene (hERG) channel. This test is being used to identify compounds that block hERG channel and prolong cardiac action potential (AP) duration (i.e. the QT interval) predisposing to Torsade de pointes tachycardia and sudden cardiac death [11]. The assessment of the torsadogenic potential of each compound in the drug discovery process also includes determination of drug's ability to prolong the AP in isolated, arterially perfused rabbit ventricular wedge preparations or canine Purkinje fibers and monitoring of heart rates and occurrence of arrhythmia in animals. Each of these endpoints has it's own specificity and sensitivity [12]. For example, hERG-expressing CHO cells lack the complexity of native CMs and cannot accurately predict the organ toxicity or lethal and arrhythmogenic side effects of compounds that block other channels or signaling pathways. Therefore, additional in vitro assays that better recapitulate human pathophysiology and diversity are needed to better predict all potential on-and off-target toxicities, reduce drug attrition rates and avoid use of animals for testing drugs that would never reach clinical application.

Pluripotent stem cells (PSCs) have unrestricted proliferation capacity, are able to differentiate into any differentiated cell type thus offering a cost-effective unlimited and invaluable source of organotypic differentiated cells relevant to assess human long-term organ toxicity. The ethical issues associated with human embryonic stem cells (ESCs) were a major concern in their application in toxicity studies. However, the Nobel prize-winning discovery that transient expression of a few transcription factors can stably convert an adult somatic cell into an early embryonic stage, i.e. into so called induced pluripotent stem cells (iPSCs) [13], has opened new possibilities in drug discovery circumventing ethical issues and problematic accessibility.

Repeated dose toxicity (RDT) occurs after repeated exposure to a substance over certain period of time. In the context of cosmetics, which are generally used for months and years, long-term RDT testing is of particular importance and forms the integral part of the quantitative risk assessment. The prediction of endpoints and hazard identification of both newly developed and existing cosmetic ingredients in humans is mainly based on the animal systems as they allow simultaneous evaluation of multiple organ systems. However, there is a great demand

and need for the development of multidisciplinary integrated approaches consisting of human in vitro models for risk assessment as an alternative to animal models as they better mimic the human in vivo system [14].

In this chapter, we will summarize the latest developments in applications of PSCs and their tissue-specific derivatives for toxicity testing. We will outline the recent developments in toxicogenomic technologies which are employed to develop and investigate human biomarkers for toxicity in PSC based models accelerating drug development process. We mainly focus on the application of PSCs in RDT testing.

2. Human pluripotent stem cells for repeated dose toxicity assessment

Human pluripotent stem cells (PSC) offer with their ability to recapitulate the most essential steps of embryonic development and give rise to different mature cell types in vitro an optimal human cellular model, which could help in increasing the safety and predictability of RDT testing leading to low late stage attrition of compounds. Combined with this cell model, toxicogenomic technologies would help predict biomarkers in an evidence-based approach.

So far, the safety assessment for novel drug candidates includes in vivo RDT tests in rodent and non-rodent models. The drawbacks of RDT studies include false negative results and unexpected humans toxicity of compounds that were judged to be safe in preclinical studies [15, 16]. Such unexpected toxicity is one of the major reasons for the withdrawal of a drug from the market. The heart and liver are often target organs in toxicology. Novel in vitro screening methods are, thus, required to classify toxic compounds earlier in development, which would lead to safer drugs, more efficient drug discovery process, lower costs and reduced laboratory animal use [17]. There is an increasing interest from biopharmaceutical industry to develop such test systems by using derivatives of human ESCs or iPSCs.

The iPSCs have a clear advantage over ESCs as they do not involve ethical issues. The generation of iPSCs involves reorganization of condensed chromatin to open state chromatin, which is aided by histone acetylation. Epigenetic factors are crucial for iPSC generation and maintenance of their pluripotent state. Although the epigenetic state of iPSCs largely resembles that of ESCs, iPSCs also have a unique DNA methylation patterns they retain epigenetic memory of the respective somatic tissue of origin which might influence their differentiation potential and affect the quality and quantity of cells for RDT [18]. On the other side, it is also well known that different agents, so called epimutagens, can cause DNA methylation and histone modification changes leading to disease [19]. These epigenetic modifications directly affect transcription factors and other chromatin binding proteins that regulate cell type-specific gene expression. The detection of biomarkers related to epigenetic modifications in RDT would be of great importance, but until now there are no systematic studies conducted in this direction. In addition, employing of iPSCs and their derivatives for this purpose poses a great challenge because genetic and epigenetic variations in iPSCs associated with reprogramming and in vitro manipulation may compromise their utility for downstream applications [20]. The lesser the variation in epigenetic changes in iPSCs the greater will be the specificity in in vitro toxicological studies. Recently, Planello and coworkers demonstrated that the choice of reprogramming factors greatly influences the DNA methylation abnormalities in iPSCs. Even highly selected iPSC lines have been shown to retain epigenetic signature of donor cell [21]. Gupta et al have shown that global transcriptional profiles of human iPSCs and ESCs are very similar and that this similarity also exists between the corresponding beating clusters derived from them [22]. They have also shown that some fibroblasts-specific mRNA expression partners were retained in the iPSCs derived from them. Significant proportion of these genes were also shown to be expressed at the same level in iPSC-derived but not in ESC-derived beating clusters indicating the retention of epigenetic memory even in the differentiated and highly enriched iPSC derivatives. Likewise, several microRNA expression profiling studies have shown the subtle differences between iPSC derivatives [23]. Hence, the iPSCs may not represent an ideal platform for RDT testing. With the current pace of iPSC research it may be possible to create iPSCs with little or no epigenetic anomalies. Polo and coworkers have shown that this retained epigenetic memory of iPSCs in early passages can be erased using extensive continued passage [24]. By using chromatin-modifying compounds like HDAC inhibitors it may be possible to stabilize the epigenetic state of iPSCs and their derivatives and decrease the frequency of heterogeneity within iPSCs. However, using the PSC-derivatives to predict RDT in human toxicological endpoints is still challenging.

3. PSC-derived cardiomyocytes for toxicity testing

Human Stringent cosmetics legislation amending directives especially within the European Union (EU) related to complete replacement of animal models in cosmetic industry safety testing by alternative methods has emphasized an urgent need for the development of reduction, refinement and replacement (3R) of the existing animal studies [25]. In order to fill gaps in non-animal alternative methods and to focus on complex RDT a research initiative called "Safety Evaluation Ultimately Replacing Animal Testing 1 (SEURAT-1)" composed of six complimentary research projects was launched in 2011 and jointly funded by the European Commission's FP7 HEALTH Programme and Cosmetics Europe (http://www.seurat-1.eu/). Embryonic Stem cell-based Novel Alternative Testing Strategies (ESNATS) is a European Union's Seventh Framework Programme (FP7), which focuses on developing a human ESC based novel toxicity test platforms to accelerate drug development. Human ESC based in vitro reproductive toxicity, neurotoxicity, toxicogenomics, proteomics and kinetics were tested for their predictive value in the identification of toxicity endpoints (http://www.esnats.eu). The RDT delivers the No Observable Adverse Effect (NOAEL), which is used in calculation of the substance safety parameters [25]. PSC-derived models hold a great potential for refinement of current models of cardiotoxicity. For many toxicology applications, a homogeneous defined population of specific cell types is required which stem cells can provide.

Validity of human PSC-derived cardiomyocytes (CM) for toxicity testing and safety pharmacology has been investigated in several studies [26-28]. The susceptibility of disease-specific human iPSC-CMs to toxicity compared to healthy human PSC-CMs was evaluated recently by Joseph Wu laboratory [29]. This group showed that disease-specific human iPSC-CM are more accurate predictors of drug-induced cardiotoxicity than standard hERG-expressing HEK293 cells. This observation suggests that human iPSC-CM may represent a suitable model for evaluation of drug safety and efficacy. However, there is still a need to examine how well the alternative systems can replace the animal models for RDT testing. The traditional repeated-dose toxicological endpoints that relate to cardiotoxicity include histopathological examinations of the heart and electrocardiographic recordings in the non-rodent species [30]. The current regulatory framework guidelines for cardiotoxicity testing include blood pressure, heart rate and electrocardiogram (ECG) parameters as well as repolarization and conductance abnormalities, cardiac output, ventricular contractility and vascular resistance. The limitations of RDT testing in vivo clearly encouraged the scientific community to identify and develop alternative in vitro methodologies to thoroughly estimate the integrated and complex responses in the endpoints that are taken into consideration.

Drugs exerting toxic effects on cardiovascular system have shown to affect the heart function in a way that includes changes in the contractility, cardiac rhythm, blood pressure and ischemia [31]. Such toxic effects have led these drugs to be withdrawn, requiring expansion of rules on cardiotoxicity testing. A new application for CMs derived from human ESCs and iPSCs has surfaced because of the lack of availability of human primary material for cardiotoxicity testing and their ability to overcome species variability. In vitro cardiotoxicity testing applications using human PSC-CMs is very advantageous and complimentary to the existing RDT applications. Endpoints such as action potential parameters, metabolic activity, membrane leakage, energy content and intracellular calcium handling can be monitored for assessing cardiotoxicity. As mentioned above, the effect of new drugs on cardiac electrophysiology (i.e. changes in ventricular repolarization) is a focus for tight control. The balanced concerted activity of several cardiac ion channels is important for proper ventricular repolarization and alterations may lead to ventricular arrhythmias. Therefore, electrophysiological assessment of the proarrhythmic potential of drugs is very relevant in cardiotoxicity assays and human PSC-derived CMs are suitable for such assays because they exhibit calcium handling properties, ion channel activity and regulatory protein expression important for the development of a mature repolarization phenotype in CMs [12].

Recently, several studies evaluated the potential of human iPSC-CMs for pharmacological screening-assays and drug discovery applications [32-34]. However, the utility of iPSC-CMs to accurately predict toxicity in humans may be limited by their immature character [35]. Current differentiation protocols give rise to heterogeneous phenotypes of spontaneously beating human PSC-CMs with structural proteins, Ca2+release units, ion channels, action potentials, and hormonal response being similar to that of native fetal CMs. However, the electrophysiological and structural properties of PSC-CMs do not fully resemble those of adult CMs. Therefore, the model based on human PSC-CMs must be improved before it can represent an ideal platform for cardiac RDT. The mentioned issues can be solved by following measures: a) by modulating cellular signaling pathways it is possible to get a homogeneous CM population [36] b) with the application of tissue engineering it is possible to create a 3D tissue constructs which provide microenvironment similar to native heart thus helping in structural maturation of CMs [37] and c) prolonged culturing of iPSC-CMs can increase the maturation of Ca2+handling [38]. Further improvements of differentiation methods will enable generation of more homogeneous and mature CM populations thus increasing their validity for RDT testing. The overall predictability of drug efficacy and toxicity using iPSC-CMs and disease specific iPSC-CMs has been recently reported by several groups [29, 39-41]. An absolute requirement for CMs to be used for RDT is to be able to stably maintain the beating cardiac phenotype for a prolonged period of time under defined conditions. Both hiPSC-CMs and hESC-CMs display beat rate variability similar to that of a human heart sinoatrial node [42]. However, recently, variability in action potentials and sodium currents in response to lidocaine and tetrodotoxin was shown in late stage in vitro differentiated human iPSC-CMs [32], thus warranting some caution and further analyses.

Bioanalytics is a very promising tool in the application of in vitro cardiotoxicty assays. Novel bioanalytical tools for discovery of biomarkers of cardiotoxicity include the field potential QT scanning, using cellular oxygen uptake for monitoring the metabolic state of CMs, using surface plasmon resonance (SPR) biosensing for key CM biomarkers, and also exploiting real-time multi-wavelength fluorimetry [12]. Novel imaging technologies and physiological analyses such as impedance measurements [43] and microelectrode arrays (MEAs) [44] give an insight into major in vitro cellular events such as migration, proliferation, cell morphology, cell–cell interactions and colony formation, relevant to biomarker discovery.

4. Stem cell-derived hepatocytes for toxicity testing

So far, hepatotoxicity is evaluated on day 28 or 90 in in vivo RDT tests by analysis of clinical parameters, hematology, and histopathology. RDT tests evaluate chronic effects on organ toxicity to establish a NOAEL which is used in calculation of the substance safety parameters [25]. The extrapolation of the quantitative risk assessment for cosmetic ingredients using data derived from animal studies to in vitro systems could be done by considering a margin of safety (MoS) value of at least 100 for intra-species and inter-species variation [25]. Human PSCs represent a promising human cellular model which could help in increasing the safety and predictability of RDT testing. Combined with this cell model, toxicogenomic technologies would help predict biomarkers in an evidence-based approach. During these RDT tests, the animals are observed for indications of toxicity. Afterwards, necropsy, blood analysis and histopathology of the organs of the animals are performed [17]. However, these parameters can turn out to be insensitive and potentially generate false negative results [15, 16]. Unexpected hepatotoxicity may be seen in the clinical trials or even when the product is already on the market because careful examinations of idiosyncratic (person specific) or non-idiosyncratic inter-drug interactions are either ignored or overseen [45]. This is also probably because of dose-dependent reactions and other unknown peculiar drug interactions. There is need for novel screening methods that can address these hepato-toxicological hazards early in the development [46]. Most studies relied on the use of liver slices as an in vitro model for toxicity testing due to limited availability of tissue samples. However, the human PSC-derived hepatocytes have the potential to replace these in vitro models and be applied for toxicity assessment. The suitability of hepatocyte-like cells derived from human PSCs for toxicity testing and drug discovery were systematically studied by several groups [47-49].

5. Omics strategies to develop biomarkers for RDT

5.1. Toxicogenomics

Proteomics, genomics and metabonomics, either alone or in combination have the potential for developing biomarkers in applied toxicology. Toxicogenomics refers to the areas mentioned above and is a thorough-mean for hi-throughput discovery of biomarkers using latest technologies [50]. Transcriptomics measures the levels of both coding and non-coding RNAs using hi-throughput technology such as microarrays. This whole genome gene expression analysis can measure the levels of expression of a gene at any stage, in any tissue and in any vitro model. Examples of toxicogenomics applications include prediction of genotoxicity or carcinogenicity, target organ toxicity and endocrine disruption. Expression profiling of any selected cellular systems exposed to new test substances is compared against controls to identify, classify and validate toxic compound and its effects. Bioinformatic analyses of the data sets obtained from above can be used to predict the patterns and signatures of a toxin (e.g. biological processes or signaling pathways affected by a toxin). Furthermore, the data sets can be matched up against existing databases for predicting and carving out a mode-of-action for the toxin. The main disadvantage of this approach is limited reproducibility and also it is semi quantitative and detects only changes in gene expression. Therefore, mRNA expression profiling cannot be used as a standalone method in identifying potential biomarkers of RDT.

EU FP7 project Predict-IV is evaluating the integration of 'omics' technologies, biomarkers and high content imaging for the early prediction of toxicity of pharmaceuticals in vitro. The aim is to identify general molecular response pathways that result from toxic drug effects that are independent of the cell/tissue type [51]. Detection of endpoints and biomarkers of RDT using in vitro systems (DETECTIVE) is a unique large scale SEURAT-1 cluster project aimed at establishing screening pipeline of high content, high throughput as well as classical functional and "-omics" technologies to detect human biomarkers for RDT in in vitro test system (http://www.detect-iv-e.eu/). Other-omics technologies such as microRNA analysis and epigenetics also play a vital role.

5.2. Proteomics

Drug induced toxicity can also exhibit various effects at the proteome level. Classification of such endpoints is difficult using traditional RDT methods. Proteomics improves the classification by identifying individual proteins or such protein panels that reflect the specific toxic pathway mechanisms. Proteomics-based in vitro toxicity assays measure drug-induced changes by comparing in vitro to in vivo effects thus validating the suitability of in vitro models. There is an absolute need for integration of standard RDT tests with the 'omics' applications. Current proteomic technologies include gel-based (1-DE or 2-DE) and gel-free (LC-MS/MS) techniques [17]. Recently thalidomide-specific proteomics signatures during human ESCs differentiation were identified using two-dimensional electrophoresis coupled with Tandem Mass spectrometry [52]. Proteomic studies are quantitative, sensitive and are more accurate and powerful in detecting protein biomarkers of RDT. Main pitfalls include posttranslational changes and limited protein detection capacity.

5.3. Metabonomics

Human PSCs offer a potential alternative test system for the identification of developmental toxicants [53]. Metabonomics refers to profiling of diverse metabolic complement of a biofluid or tissue using analytical tools such as high-field NMR together with mass spectrometry [54]. Subsequent statistical modeling and analysis of a multivariate spectral profiles obtained using NMR [55] in combination with LC-MS and UPLC helps to distinguish the phenotypes and metabolites of interest. These metabolites might represent new biomarkers for toxicity. Previously, some of metabolites were identified to be biomarkers for a variety of pathological diseases [56]. Metabonomics-based approaches have proved to be highly successful in furthering our understanding of research in the field of drug metabolism, drug pathways and toxicology [54, 55]. In addition, metabonomics provides a useful link between 'omics' platforms such as genomics, transcriptomics and proteomics and end-stage histopathological analyses [54].

The Consortium for Metabonomic Toxicology (COMET) project (a collaboration between five pharmaceutical companies and Imperial College London) focused on pre-clinical toxicological research and resulted in the generation of an extensive 1H NMR biofluid spectral database which was used for screening of toxins and also to build an expert system for prediction of target organ toxicity [57]. A follow-up project, COMET-2, is currently investigating the detailed biochemical mechanisms of toxicity, and seeks a better understanding of inter-subject variation in metabonomics analyses [55]. Several groups have developed Metabonomics-based robust human ESC in vitro test systems for predicting human developmental toxicity biomarkers and pathways [58, 59]. Metabonomics is the most relevant and robust omics platform to study both in vivo and in vitro toxicology. It is possible to detect metabolites with accuracy but it is limited by its high costs and complex metabolite isolation procedures.

6. Bioinformatic and statistical analysis of candidate biomarkers

Since the omics methods are extensively data-intensive and bulk, there is a definite need for bioinformatics and statistical analysis for organizing the data in conveniently accessible databases, which integrate huge number of data sets, and therefore need quality database manager software such as SQL for centralized storage and flexible web based access to the bulk data.

Noncommercial databases available on the web such as CEBS (chemical effects in biological systems, (http://cebs.niehs.nih.gov), PhenoGen (http://phenogen.uchsc.edu), along with and commercial databases like ArrayTrack and ArrayExpress (http://www.ebi.ac.uk/arrayexpress) help to generate large data sets. These are complemented by metabonomics databases (http://

www.hmdb.ca). PrestOMIC is proteome-specific open-source that is a user-friendly database, where researchers can upload and share data with the scientific community using a customizable browser [60]. Such a database helps researchers to increase the exposure and impact of their data by enabling extensive data set comparisons.

Another open-source systems biology application called SysBio-OM, integrates information from the CEBS database with other open source projects, including MAGE-OM (micro-array gene expression object model) and PEDRo (proteomics experiment data repository), to model profiling of protein, and metabolite expression and protein-protein interactions following insult [61]. SysTox-OM is a more specific application that performs expression profiling of genome, proteome, and metabolome, after the introduction of a toxicant. Different omics approaches and some of the crucial data bases are summarized in Table.1. While incorporating toxicological endpoints such as – clinical chemistry, hematology, observations and histopathology, it profiles the phenotype [62]. With this application, one can identify a single toxic phenotype, classify, and compare gene and protein expression profiles in an organ after administration of each drug. It is also possible to predict a common toxicologic pathway, mechanism or a biomarker. ToxBank is an EU FP7 project aimed at establishing a dedicated web-based warehouse for toxicity data management and modeling along with establishing a cell and tissue banking information for in vitro toxicity testing.

	Databases	Web links
Toxicogenomics	Comparative Toxicogenomics Database (CTD)	http://ctdbase.org/
	Open TG-GATEs	http://toxico.nibio.go.jp/english/
	DrugMatrix®	https://ntp.niehs.nih.gov/drugmatrix/index.html
	TOXNET	http://toxnet.nlm.nih.gov/index.html
	ArrayExpress	http://www.ebi.ac.uk/arrayexpress/
	diXa Data Warehouse	http://wwwdev.ebi.ac.uk/fg/dixa/index.html
	ArrayTrack™	http://www.fda.gov/ScienceResearch/ BioinformaticsTools/Arraytrack/default.htm
Proteomics	PRIDE	http://www.ebi.ac.uk/pride/archive/
	ProteomicsDB	https://www.proteomicsdb.org/
	GPMDB	http://gpmdb.thegpm.org/
	NIST	http://peptide.nist.gov/
Metabonomics	Human Metabolome Database (HMDB)	http://www.hmdb.ca/
	BiGG	http://bigg.ucsd.edu/
	MetaboLights	http://www.ebi.ac.uk/metabolights/index
	MMCD	http://fiehnlab.ucdavis.edu/projects/binbase_setupx
	SetupX & BinBase	http://www.bml-nmr.org/
	BML-NMR	http://www.massbank.jp/
	MassBank	http://gmd.mpimp-golm.mpg.de/
	Golm Metabolome Database (GMD)	

Table 1. Summary of different Omics approaches and corresponding databases

7. Conclusion and future perspective

Improved toxicity testing methods complementing advanced in vitro assays are very crucial in reducing the rate of attrition in final stages of product development. To avoid failures and withdrawals, there is an absolute need for integration of all available technologies to minimize cumbersome process of trials and expenses and eventually reduce the increasing costs of bringing a new drug into market. Supplementing toxicology evaluation methods, such as histopathology, physiology and clinical chemistry with transcriptomics, proteomics and metabonomics could provide new insights into the mechanisms underlying toxicological pathologies. Integration of in vitro toxicology technologies, with systems biology methods resulted in 'systems toxicology'. Expansion of open source databases and analytical platforms is critical to the discovery of novel biomarkers of toxicity. So far, the available approaches for discovery of biomarkers included toxicogenomics, toxicoproteomics, metabonomics and bioinformatics analyses (systems biology approach) while the technologies available for quantification include ELISA, solid phase ELISA, Luminex technology and patterned paper technology [50]. Individual technologies have limited usefulness unless the data generated from these assay platforms and '-omics' discovery technologies are integrated. The discovery of DNA microarrays and protein chips has made information exchanges extraordinarily easy, convenient and quick. Integration of information from these powerful sources using analytical computing software products, noncommercial databases, and advances in hi-throughput technology is the future of the next phase in the identification, selection and qualification of novel biomarkers of toxicity.

Future of toxicogenomics lies in developing a more refined understanding of molecular mechanisms related to specific toxicologies, to elucidate molecular signatures associated with the prediction of biomarkers or panels of biomarkers with support from the field of transcriptomics, metabolomics, and proteomics. These analytical tools applied to the emerging human PSC-based in vitro platforms utilizing their organ-specific differentiated derivatives, such as CMs, hepatocytes and neurons, have a great potential to revolutionize the field of toxicology. However, the full potential of these human in vitro cell-based platforms in predicting toxicity of compounds in humans will be realized only with further improvements in derivation of highly standardized, well-defined and homogeneous cell populations that functionally and structurally strongly resemble their adult counterparts and development of sensitive and robust methods for accurate detection of toxicity.

Acknowledgements

The work in the author's laboratories is supported by the European Union FP7 Program, Bundesministerium für Bildung und Forschung (BMBF), Else-Kröner-Fresenius Stiftung, Excellence Research Support Program of the University of Cologne and Köln Fortune Program.

http://dx.doi.org/10.5772/58485

Author details

Shiva Prasad Potta[2], Tomo Šarić[1], Michael Heke[1], Harinath Bahudhanapati[2] and Jürgen Hescheler[1*]

*Address all correspondence to: J.Hescheler@uni-koeln.de

1 Center for Physiology and Pathophysiology, Institute for Neurophysiology, University of Cologne, Cologne, Germany

2 Tulip Bio-Med Solutions (P) Ltd., Hyderabad, Andhra Pradesh, India

References

[1] Kola I, Landis J. Can the pharmaceutical industry reduce attrition rates? Nature reviews Drug discovery. 2004;3(8):711-5.

[2] Chapman KL, Holzgrefe H, Black LE, Brown M, Chellman G, Copeman C, et al. Pharmaceutical toxicology: designing studies to reduce animal use, while maximizing human translation. Regulatory toxicology and pharmacology : RTP. 2013;66(1): 88-103.

[3] Hutchinson L, Kirk R. High drug attrition rates--where are we going wrong? Nature reviews Clinical oncology. 2011;8(4):189-90.

[4] Force T, Kolaja KL. Cardiotoxicity of kinase inhibitors: the prediction and translation of preclinical models to clinical outcomes. Nature reviews Drug discovery. 2011;10(2):111-26.

[5] Kannankeril PJ, Roden DM. Drug-induced long QT and torsade de pointes: recent advances. Current opinion in cardiology. 2007;22(1):39-43.

[6] Mordwinkin NM, Burridge PW, Wu JC. A review of human pluripotent stem cell-derived cardiomyocytes for high-throughput drug discovery, cardiotoxicity screening, and publication standards. Journal of cardiovascular translational research. 2013;6(1): 22-30.

[7] Collins BC, Sposny A, McCarthy D, Brandenburg A, Woodbury R, Pennington SR, et al. Use of SELDI MS to discover and identify potential biomarkers of toxicity in InnoMed PredTox: a multi-site, multi-compound study. Proteomics. 2010;10(8): 1592-608.

[8] Kaese S, Verheule S. Cardiac electrophysiology in mice: a matter of size. Frontiers in physiology. 2012;3:345.

[9] Roden DM, Altman RB, Benowitz NL, Flockhart DA, Giacomini KM, Johnson JA, et al. Pharmacogenomics: challenges and opportunities. Annals of internal medicine. 2006;145(10):749-57.

[10] McKim JM, Jr. Building a tiered approach to in vitro predictive toxicity screening: a focus on assays with in vivo relevance. Combinatorial chemistry & high throughput screening. 2010;13(2):188-206.

[11] Pollard CE, Valentin JP, Hammond TG. Strategies to reduce the risk of drug-induced QT interval prol.ongation: a pharmaceutical company perspective. British journal of pharmacology. 2008;154(7):1538-43.

[12] Mandenius CF, Steel D, Noor F, Meyer T, Heinzle E, Asp J, et al. Cardiotoxicity testing using pluripotent stem cell-derived human cardiomyocytes and state-of-the-art bioanalytics: a review. Journal of applied toxicology : JAT. 2011;31(3):191-205.

[13] Takahashi K, Tanabe K, Ohnuki M, Narita M, Ichisaka T, Tomoda K, et al. Induction of pluripotent stem cells from adult human fibroblasts by defined factors. Cell. 2007;131(5):861-72.

[14] Adler S, Basketter D, Creton S, Pelkonen O, van Benthem J, Zuang V, et al. Alternative (non-animal) methods for cosmetics testing: current status and future prospects-2010. Archives of toxicology. 2011;85(5):367-485.

[15] Olson H, Betton G, Stritar J, Robinson D. The predictivity of the toxicity of pharmaceuticals in humans from animal data--an interim assessment. Toxicology letters. 1998;102-103:535-8.

[16] Suter L, Schroeder S, Meyer K, Gautier JC, Amberg A, Wendt M, et al. EU framework 6 project: predictive toxicology (PredTox)--overview and outcome. Toxicology and applied pharmacology. 2011;252(2):73-84.

[17] Van Summeren A, Renes J, van Delft JH, Kleinjans JC, Mariman EC. Proteomics in the search for mechanisms and biomarkers of drug-induced hepatotoxicity. Toxicology in vitro : an international journal published in association with BIBRA. 2012;26(3): 373-85.

[18] Bar-Nur O, Russ HA, Efrat S, Benvenisty N. Epigenetic memory and preferential lineage-specific differentiation in induced pluripotent stem cells derived from human pancreatic islet beta cells. Cell stem cell. 2011;9(1):17-23.

[19] Szyf M. Epigenetics, DNA methylation, and chromatin modifying drugs. Annual review of pharmacology and toxicology. 2009;49:243-63.

[20] Liang G, Zhang Y. Genetic and epigenetic variations in iPSCs: potential causes and implications for application. Cell stem cell. 2013;13(2):149-59.

[21] Kim K, Doi A, Wen B, Ng K, Zhao R, Cahan P, et al. Epigenetic memory in induced pluripotent stem cells. Nature. 2010;467(7313):285-90.

[22] Gupta MK, Illich DJ, Gaarz A, Matzkies M, Nguemo F, Pfannkuche K, et al. Global transcriptional profiles of beating clusters derived from human induced pluripotent stem cells and embryonic stem cells are highly similar. BMC developmental biology. 2010;10:98.

[23] Wilson KD, Venkatasubrahmanyam S, Jia F, Sun N, Butte AJ, Wu JC. MicroRNA profiling of human-induced pluripotent stem cells. Stem cells and development. 2009;18(5):749-58.

[24] Polo JM, Liu S, Figueroa ME, Kulalert W, Eminli S, Tan KY, et al. Cell type of origin influences the molecular and functional properties of mouse induced pluripotent stem cells. Nature biotechnology. 2010;28(8):848-55.

[25] Vanhaecke T, Pauwels M, Vinken M, Ceelen L, Rogiers V. Towards an integrated in vitro strategy for repeated dose toxicity testing. Archives of toxicology. 2011;85(5): 365-6.

[26] Peng S, Lacerda AE, Kirsch GE, Brown AM, Bruening-Wright A. The action potential and comparative pharmacology of stem cell-derived human cardiomyocytes. Journal of pharmacological and toxicological methods. 2010;61(3):277-86.

[27] Jonsson MK, Duker G, Tropp C, Andersson B, Sartipy P, Vos MA, et al. Quantified proarrhythmic potential of selected human embryonic stem cell-derived cardiomyocytes. Stem cell research. 2010;4(3):189-200.

[28] Guo L, Abrams RM, Babiarz JE, Cohen JD, Kameoka S, Sanders MJ, et al. Estimating the risk of drug-induced proarrhythmia using human induced pluripotent stem cell-derived cardiomyocytes. Toxicological sciences : an official journal of the Society of Toxicology. 2011;123(1):281-9.

[29] Liang P, Lan F, Lee AS, Gong T, Sanchez-Freire V, Wang Y, et al. Drug screening using a library of human induced pluripotent stem cell-derived cardiomyocytes reveals disease-specific patterns of cardiotoxicity. Circulation. 2013;127(16):1677-91.

[30] Misner DL, Frantz C, Guo L, Gralinski MR, Senese PB, Ly J, et al. Investigation of mechanism of drug-induced cardiac injury and torsades de pointes in cynomolgus monkeys. British journal of pharmacology. 2012;165(8):2771-86.

[31] Brana I, Tabernero J. Cardiotoxicity. Annals of oncology : official journal of the European Society for Medical Oncology / ESMO. 2010;21 Suppl 7:vii173-9.

[32] Sheng X, Reppel M, Nguemo F, Mohammad FI, Kuzmenkin A, Hescheler J, et al. Human pluripotent stem cell-derived cardiomyocytes: response to TTX and lidocain reveals strong cell to cell variability. PloS one. 2012;7(9):e45963.

[33] Sirenko O, Crittenden C, Callamaras N, Hesley J, Chen YW, Funes C, et al. Multiparameter in vitro assessment of compound effects on cardiomyocyte physiology using iPSC cells. Journal of biomolecular screening. 2013;18(1):39-53.

[34] Harris K, Aylott M, Cui Y, Louttit JB, McMahon NC, Sridhar A. Comparison of electrophysiological data from human-induced pluripotent stem cell-derived cardiomyocytes to functional preclinical safety assays. Toxicological sciences : an official journal of the Society of Toxicology. 2013;134(2):412-26.

[35] Yang X, Pabon L, Murry CE. Engineering adolescence: maturation of human pluripotent stem cell-derived cardiomyocytes. Circulation research. 2014;114(3):511-23.

[36] Zhang Q, Jiang J, Han P, Yuan Q, Zhang J, Zhang X, et al. Direct differentiation of atrial and ventricular myocytes from human embryonic stem cells by alternating retinoid signals. Cell research. 2011;21(4):579-87.

[37] Turnbull IC, Karakikes I, Serrao GW, Backeris P, Lee JJ, Xie C, et al. Advancing functional engineered cardiac tissues toward a preclinical model of human myocardium. FASEB journal : official publication of the Federation of American Societies for Experimental Biology. 2014;28(2):644-54.

[38] Lundy SD, Zhu WZ, Regnier M, Laflamme MA. Structural and functional maturation of cardiomyocytes derived from human pluripotent stem cells. Stem cells and development. 2013;22(14):1991-2002.

[39] Dick E, Rajamohan D, Ronksley J, Denning C. Evaluating the utility of cardiomyocytes from human pluripotent stem cells for drug screening. Biochemical Society transactions. 2010;38(4):1037-45.

[40] Braam SR, Tertoolen L, van de Stolpe A, Meyer T, Passier R, Mummery CL. Prediction of drug-induced cardiotoxicity using human embryonic stem cell-derived cardiomyocytes. Stem cell research. 2010;4(2):107-16.

[41] Yokoo N, Baba S, Kaichi S, Niwa A, Mima T, Doi H, et al. The effects of cardioactive drugs on cardiomyocytes derived from human induced pluripotent stem cells. Biochemical and biophysical research communications. 2009;387(3):482-8.

[42] Binah O, Weissman A, Itskovitz-Eldor J, Rosen MR. Integrating beat rate variability: from single cells to hearts. Heart rhythm : the official journal of the Heart Rhythm Society. 2013;10(6):928-32.

[43] Nguemo F, Saric T, Pfannkuche K, Watzele M, Reppel M, Hescheler J. In vitro model for assessing arrhythmogenic properties of drugs based on high-resolution impedance measurements. Cellular physiology and biochemistry : international journal of experimental cellular physiology, biochemistry, and pharmacology. 2012;29(5-6): 819-32.

[44] Reppel M, Boettinger C, Hescheler J. Beta-adrenergic and muscarinic modulation of human embryonic stem cell-derived cardiomyocytes. Cellular physiology and biochemistry : international journal of experimental cellular physiology, biochemistry, and pharmacology. 2004;14(4-6):187-96.

[45] Hartung T. Toxicology for the twenty-first century. Nature. 2009;460(7252):208-12.

[46] Amacher DE. The discovery and development of proteomic safety biomarkers for the detection of drug-induced liver toxicity. Toxicology and applied pharmacology. 2010;245(1):134-42.

[47] Takayama K, Kawabata K, Nagamoto Y, Kishimoto K, Tashiro K, Sakurai F, et al. 3D spheroid culture of hESC/hiPSC-derived hepatocyte-like cells for drug toxicity testing. Biomaterials. 2013;34(7):1781-9.

[48] Ulvestad M, Nordell P, Asplund A, Rehnstrom M, Jacobsson S, Holmgren G, et al. Drug metabolizing enzyme and transporter protein profiles of hepatocytes derived from human embryonic and induced pluripotent stem cells. Biochemical pharmacology. 2013;86(5):691-702.

[49] Yildirimman R, Brolen G, Vilardell M, Eriksson G, Synnergren J, Gmuender H, et al. Human embryonic stem cell derived hepatocyte-like cells as a tool for in vitro hazard assessment of chemical carcinogenicity. Toxicological sciences : an official journal of the Society of Toxicology. 2011;124(2):278-90.

[50] Collings FB, Vaidya VS. Novel technologies for the discovery and quantitation of biomarkers of toxicity. Toxicology. 2008;245(3):167-74.

[51] Adler S, Lindqvist J, Uddenberg K, Hyllner J, Strehl R. Testing potential developmental toxicants with a cytotoxicity assay based on human embryonic stem cells. Alternatives to laboratory animals : ATLA. 2008;36(2):129-40.

[52] Meganathan K, Jagtap S, Wagh V, Winkler J, Gaspar JA, Hildebrand D, et al. Identification of thalidomide-specific transcriptomics and proteomics signatures during differentiation of human embryonic stem cells. PloS one. 2012;7(8):e44228.

[53] Chapin RE, Stedman DB. Endless possibilities: stem cells and the vision for toxicology testing in the 21st century. Toxicological sciences : an official journal of the Society of Toxicology. 2009;112(1):17-22.

[54] Coen M. A metabonomic approach for mechanistic exploration of pre-clinical toxicology. Toxicology. 2010;278(3):326-40.

[55] Nicholson JK, Lindon JC. Systems biology: Metabonomics. Nature. 2008;455(7216): 1054-6.

[56] Holmes E, Wilson ID, Nicholson JK. Metabolic phenotyping in health and disease. Cell. 2008;134(5):714-7.

[57] Ebbels TM, Keun HC, Beckonert OP, Bollard ME, Lindon JC, Holmes E, et al. Prediction and classification of drug toxicity using probabilistic modeling of temporal metabolic data: the consortium on metabonomic toxicology screening approach. Journal of proteome research. 2007;6(11):4407-22.

[58] West PR, Weir AM, Smith AM, Donley EL, Cezar GG. Predicting human developmental toxicity of pharmaceuticals using human embryonic stem cells and metabolomics. Toxicology and applied pharmacology. 2010;247(1):18-27.

[59] Kleinstreuer NC, Smith AM, West PR, Conard KR, Fontaine BR, Weir-Hauptman AM, et al. Identifying developmental toxicity pathways for a subset of ToxCast chemicals using human embryonic stem cells and metabolomics. Toxicology and applied pharmacology. 2011;257(1):111-21.

[60] Howes CG, Foster LJ. PrestOMIC, an open source application for dissemination of proteomic datasets by individual laboratories. Proteome science. 2007;5:8.

[61] Xirasagar S, Gustafson S, Merrick BA, Tomer KB, Stasiewicz S, Chan DD, et al. CEBS object model for systems biology data, SysBio-OM. Bioinformatics. 2004;20(13): 2004-15.

[62] Xirasagar S, Gustafson SF, Huang CC, Pan Q, Fostel J, Boyer P, et al. Chemical effects in biological systems (CEBS) object model for toxicology data, SysTox-OM: design and application. Bioinformatics. 2006;22(7):874-82.

Chapter 9

The Minipig — A New Tool in Stem Cell Research

Hideyuki Kobayashi, Toshihiro Tai,
Koichi Nagao and Koichi Nakajima

Additional information is available at the end of the chapter

http://dx.doi.org/10.5772/57603

1. Introduction

The establishment of pluripotent mouse embryonic stem cells (ESCs), which have the capacity to differentiate into all cell types in the mammalian body, was first described by Evans and Kaufman in 1981. Seventeen years later, in 1998, the first report of human ESCs, with similar differentiation characteristics was published by James Thompson. In 2006, Takahashi and Yamanaka discovered that the genome of a differentiated somatic cell could be epigenetically reprogrammed to pluripotency by inducing the expression of pluripotency transcription factors, resulting in the production of pluripotent stem cells (iPSCs). This research has accelerated the study of regenerative medicine.

In 2009, the US Food and Drug Administration approved the first Phase I clinical trials of human ESC-derived cells for the treatment of spinal cord injuries. However, previous reports had revealed that stem cell-based therapies increase the risk of tumor development. Therefore, further basic research is needed before stem cell-based therapies can be applied in the clinic. The development of stem cell tools can be critically evaluated using animal models that express human disease genes.

Research on the stem cell biology of the minipig is developing rapidly. Although research on mouse and human stem cells currently predominates over that in other species, data from these well-studied species have provided a good foundation for current and future porcine stem cell research. In addition, the increasing popularity of alternative-species models for the study of human diseases and disease mechanisms has further spurred porcine stem cell research. As a model system for pluripotent embryonic stem cell research, however, the pig presents several challenges as compared with mice and humans. Nonetheless, porcine minipig embryonic germ cells have recently been produced, and may prove particularly useful for *in vitro* and *in vivo* differentiation studies, gene targeting, and the creation of transgenic animals. In addition, studies involving the transplantation of somatic mesenchymal stem cells into porcine heart, cartilage, and bone have yielded very promising results. Lastly, minipig induced pluripotent stem cells have been established by using Sendai viruses to introduce pluripotency transcription factors into the cells.

Thus, despite the challenges of developing porcine pluripotent stem cells, recent successes in the fields of both induced pluripotent stem cells and somatic stem cells suggest that the future of research using minipig stem cells is quite promising.

2. Embryonic stem cells

Most porcine embryonic stem cell research has been performed in large domestic pig breeds rather than minipigs. Therefore, this discussion will focus on studies performed on the large domestic breeds, yet with attention paid to outcomes observed in the minipig. Research in the domestic species is highly transferrable to the minipig.

Although researchers have sought to establish ESCs in the pig, the characterization of ESCs in this animal falls short, due to a lack of both *in vivo* and long-term culture studies as compared to ESCs of mouse or human origin. The first reports on ESC production in the pig were in 1990 [1] [2] [3] [4]. However, those attempts produced only putative ESCs or embryonic stem-like cells.

A number of standard techniques are typically employed to verify the identity of true ESCs. Several *in vitro* techniques are generally performed to determine the expression profile of the ESCs, including gene-expression and protein analyses. The transcription factors expressed in both mouse and human ESCs include OCT4, NANOG, and SOX2. In addition, stage-specific embryonic antigen 1 (SSEA1) is expressed in mouse but not human ESCs, and SSEA3 and SSEA4 are detected in human but not mouse ESCs [5].

While the efforts to establish porcine ESCs have been well reviewed in the scientific literature [6] [7], a general overview of the research suggests that true porcine ESCs have not yet been produced. The production of porcine epiblast stem cells (EpiSCs) was recently reported [8]. These cells are thought to be derived from the epiblast, rather than the inner cell mass, of the developing embryo. Evidence suggests that mouse ESCs are of inner cell mass origin, while human ESCs originate from epiblasts [9]. The porcine epiblast stem cells could be cultured for 22 passages, and could differentiate *in vitro* into cell types representative of the three embryonic germ lineages as well as germ precursor cells and trophectoderm. However, it is not known whether these cells also demonstrate pluripotency.

To date, two research groups have attempted to produce porcine ESCs from the minipig [10, 11]. Li and colleagues reported that outgrowth cultures could be obtained after isolating the inner cell mass from Chinese minipig blastocysts, although sustained culture was difficult to achieve, and only a preliminary characterization of these cells was performed [10]. Long-term cultures of porcine ESC-like cells were reported by Kim and colleagues [11]. In this study, porcine ESC-like cells were derived from cloned blastocysts. These embryos were produced

http://dx.doi.org/10.5772/57603

by somatic cell nuclear transfer, using minipig fetal or neonatal fibroblasts as the donor and prepubertal gilt oocytes, followed by culture *in vitro* to the blastocyst stage of development. Two cells lines could be cultured for more than 48 passages, and expressed alkaline phosphatase (AP), SSEA1 and SSEA3, OCT4, TRA-1-60, and TRA-1-81. The only method used to characterize their differentiation ability was the observation of spontaneous differentiation into embryoid bodies (EBs), which are spheres of cells that contain cell types of all three germ layers. These EBs were assessed solely by gene analysis, thus further characterization would be necessary before they could be verified as true ESCs.

Only two reports to date describe the production of chimeras from porcine ESC-like cells: one in 1999, from Chen and colleagues [12], and a more recent publication in 2010 [13]. The first report claimed that somatic chimeric piglets could be produced, although clear analyses of these chimeric animals was lacking. The second report indicated that porcine ESC-like cells from an early passage could form chimeric piglets. The chimeric contribution was low, however; only 4 chimeric piglets were born after the transfer of hundreds of embryos. Only 2 of the 4 chimeric piglets showed coat chimerism, and this contribution was low and restricted to a single spot near the tail. Such a chimeric contribution is much lower than what would be expected from mouse ESCs, indicating that improvements to the cell culture conditions may be required to improve the plasticity of these cells.

There are several possible reasons for the difficulty in producing porcine ESCs. The lack of defined culture conditions may be one reason. Pluripotency appears to be controlled by more than one cell-signaling pathway, and these pathways are different in mouse and human ESC lines. The origin of the cells, that is, the inner cell mass (mouse) versus the epiblasts (human) may contribute to this diversity of regulatory pathways. This idea is supported by a recent publication showing that even mouse EpiSCs regulate pluripotency slightly differently from human ESCs [14]. The cell signaling that governs pluripotency in the pig remains largely unknown, although the details are beginning to be investigated [15]. For example, it was reported that fibroblast growth factor (FGF) signaling may be active in porcine epiblasts, and that the JAK/STAT pathway is inactive. The Activin/Nodal pathway also appears to be active in porcine epiblasts [8]. Culturing porcine epiblasts in medium containing basic FGF (bFGF) cannot prevent their differentiation, indicating that other factors are apparently necessary to help maintain cellular pluripotency.

Differences occurring during early embryonic development in the domestic pig, as compared with mouse and human, could also account for the observed difficulties in producing porcine ESCs. The early development of the porcine embryo prior to implantation takes longer and is less advanced than in the other species. The inner cell mass differentiates into the hypoblast and epiblast at a later time point than in the mouse and human, and the porcine epiblast expands and develops over a period of several days. The cell signaling controlling this development in the pig could differ markedly from that in the mouse or human, and should be investigated to ascertain which stage of development is optional for isolating the pluripotent cells. It is possible that the later epiblast is already predetermined at the cell-signaling level to undergo gastrulation, or that the inner cell mass cells have not yet acquired the necessary cell signaling machinery to support proliferation.

We have yet to observe *bona fide* ESCs produced from the pig. Nevertheless, continued research toward this goal is important, as the pig is a particularly useful biomedical model for studying human disease, and ESCs are a unique cell type that is especially useful for studying human disease.

3. Embryonic germ cells

Embryonic germ cells (EGCs) are pluripotent cells derived *in vitro* from primordial germ cells (PGCs). They were first derived in the mouse by Matsui et al. [16] and Resnick et al. [17]. Mouse EGCs are morphologically similar to mouse ESCs, and express similar pluripotency markers such as AP, SSEA1, Oct4, Sox2, and Nanog. The differentiation potential of these cells is also similar to that of mouse ESCs, as evidenced by their ability to differentiate *in vitro* into different tissues from the three primary germ layers [18], as well as to re-enter the germline when injected into blastocyst embryos [19]. Mouse EGCs may thus be considered equivalent to ESCs.

Similar to the conditions required to derive mouse ESCs, mouse EGCs need two important growth factors in addition to leukemia inhibitory factor (LIF) for their successful derivation from PGCs. The first is stem cell factor (SCF), a ligand for c-Kit, which supports PGC survival *in vivo* and *in vitro* by suppressing apoptosis [20]. The second is bFGF, which plays a major role in the reprogramming of mouse PGCs into pluripotent cells *in vitro* [21]. Although the mechanism of this reprogramming is still poorly understood, bFGF is known to downregulate the expression of basic lymphocyte maturation protein 1 (Blimp1) *in vitro*, which in turn causes the upregulation of c-Myc and Klf4 [21]. Because mouse PGCs also express Oct4 and Sox2, a similar mechanism may be involved in their reprogramming to ESCs.

Porcine EGCs were first derived by Shim et al. [22], who used embryos from domestic breeds on day 25 of gestation under conditions similar to those used to culture mouse EGCs. The established cell lines had an ESC-like morphology, expressed AP, and were able to differentiate into different cell types *in vitro* and *in vivo*. Other groups have derived porcine EGC lines using day 25-28 embryos from domestic breeds [23] [24] [25] [26] [27], and from the Chinese minipig [28]. Recently, putative porcine EGC lines derived from the PGCs of day 20-24 embryos from Danish Landrace crosses and Yucatan minipigs were reported [29]. Notably, these findings suggest that EGC lines can be established from any porcine breed, unlike mouse ESCs, which are restricted to certain strains.

Immunocytochemical analysis of pluripotency marker expression showed that, in addition to AP activity, EGCs express Oct4, SSEA1, SSEA3, and SSEA4 at variable levels. Analysis of the porcine EGC gene expression suggested that they are similar to human EGCs in expressing AP, Oct4, SSEA1, and SSEA4 [30]. The pig EGCs typically have long cell cycles, and proliferate slowly over many passages.

The tissue culture conditions used by all groups for the derivation and propagation of porcine EGCs are similar to those used for human and mouse ESCs. The cells are grown on feeder layers of mitotically inactivated embryonic mouse fibroblasts, most often on immortalized mouse fibroblast cells from a cell line known as STO. The initial establishment of cell lines is hampered by the fact that large numbers of porcine PGCs undergo apoptosis and die in culture within 6 hours of their incubation [25]. However, by using protease inhibitors and antioxidants to reduce the level of PGC apoptosis *in vitro*, Lee et al. were able to increase the number of EGC colonies in the primary culture. In addition, alpha 2-macroglobulin together with N-acetyl-cystein and butylated hydroxyanisole (both antioxidants) increased the number of AP-positive colonies in primary cultures at least twofold. Another approach is to use growth factors and various feeder cells to increase the survival and prevent the differentiation of PGCs in long-term culture. Lee et al. [26] showed that supplementing the culture medium with three growth factors (LIF, SCF, and bFGF) increased the number of colonies obtained in primary culture, and improved the quality of the colonies in subsequent passages compared to cells treated with only two of these growth factors. The use of feeder layers other than STO cells did not have a significant effect on PGC survival or on the quality of the derived EGCs. In another study, membrane-bound SCF and soluble LIF were sufficient to increase the number of surviving PGCs on days 3 and 5 of primary culture, while adding bFGF did not affect the results significantly [23]. It is possible that the membrane-bound form of SCF is more potent than the soluble form, which could explain these differing results. However, it has also been reported that established PGCs are capable of generating chimeras without added growth factors, while adding LIF to the culture does not improve the efficiency of establishing EGC lines [22]. Thus, further study is needed to elucidate the effects of these supplements in culture.

Pluripotent stem cells are able to form EBs, spherical aggregates containing differentiated cells, when cultured in suspension. Porcine EGCs cultured in "hanging drops" can also form simple EBs, consisting of large epithelial-like cells on the periphery surrounding mesenchymal-like cells in the center. When allowed to attach to gelatin-treated plastic dishes, the EB cells proliferate and spread on the surface of the dish, giving rise to several different types of cells [27].

Pluripotent cells can proliferate and differentiate *in vivo*, forming tumors containing differentiated tissues called teratomas, when injected into immunodeficient mice. To date, only one group has reported teratoma testing for porcine EGCs, using cell lines derived from the Chinese minipig [28]. The authors reported that the teratomas contained cells from the three primary germ layers: epithelial, neuroepithelial, and adipose tissue. In contrast to these results, two studies showed that the injection of human EGCs into immunodeficient mice failed to generate teratomas. [31] [32]. However, more recently, the formation of teratomas from human EGCs cultured under serum-free conditions was reported [33].

Another way of testing the differentiation potential of pluripotent cells *in vivo* is by chimera formation, in which the cells injected into early embryos contribute to the three germ layers and potentially to the germline. Unlike mouse ESCs and EGCs, which have been shown to re-enter the germline of chimeras, pig EGC chimeras have not displayed a proven germline contribution; furthermore, the somatic tissues of these porcine chimeras contained only a low percentage of donor-derived cells [22] [34] [24]. Other researchers demonstrated a similarly low chimeric contribution after the injection of somatic cells into sheep blastocysts and 8-cell mouse embryos [35, 36]. This is troubling because germline contribution is considered to be the ultimate proof of pluripotency, and has been demonstrated only in the mouse and chicken to date. Human ESCs are presumed to be true stem cells as well, even though their germline potential has not been tested for ethical reasons. Caution is needed when interpreting chimera experimental results. Proper controls are necessary to distinguish the "true" stem cells from somatic cells that can be integrated into the embryos after partial reprogramming by the surrounding embryonic cells.

The reasons for the limited pluripotency of the porcine EGCs are currently unclear. It is possible that the tissue culture conditions are not sufficient for maintaining the pluripotency of the PGCs *in vitro*, similar to the problems that exist in the cultures of inner cell mass and epiblast cells. Another possibility is that the pig PGCs do not undergo full epigenetic reprogramming into pluripotency like the mouse PGCs do. One indication that there might be differences in the biology of PGCs between the pig and mouse is that porcine PGCs survive and proliferate *in vitro* in the absence of externally added growth factors, regardless of whether they are cultured in serum-supplemented or serum-free conditions, while mouse PGCs fail to survive in the absence of any of the three growth factors LIF, SCF, or bFGF. In addition, mouse PGCs have not been shown to form chimeras, whereas porcine PGCs can contribute to somatic tissues after their injection into early blastocysts [24]. Thus, some of the molecular mechanisms that are important for reprogramming mouse PGCs to pluripotent EGCs in culture may be different in the pig.

4. Induced pluripotent stem cells

Takahashi and Yamanaka [37] discovered that the genome of a differentiated somatic cell can be epigenetically reprogrammed to pluripotency by the induced expression of pluripotency transcription factors, resulting in the generation of pluripotent stem cells [induced Pluripotent Stem Cells (iPSCs)]. These authors initially expressed 24 pluripotency genes in fetal fibroblasts to reprogram them into pluripotent cells, and subsequently found that the expression of only 4 of these genes, namely *Oct4, Sox2, c-myc*, and *Klf4*, was sufficient to achieve the same results. Nanog was dispensable for the cellular reprogramming. In addition, a year later, Okita et al. [38] produced germline-competent mouse iPSCs using the same growth factors, but by selecting the reprogrammed cells by Nanog expression.

The production of human iPSCs started shortly after the publication of the first report of mouse iPSC derivation. Thus, it was not surprising that the establishment of the first human iPSCs was reported by two groups simultaneously. The laboratory that produced the first mouse iPSCs successfully generated iPSCs from adult human fibroblasts by using the same 4 transcription factors employed in the initial experiment [39]. The other group, from the laboratory of human ESC pioneer James Thompson, was able to reprogram human fibroblasts by expressing OCT4, SOX2, NANOG, and LIN28 [40].

Predictably, these studies sparked significant interest among researchers working in the area of porcine pluripotent cells. Within two years after the publication of the first papers describing human and mouse iPSCs, three different groups reported the establishment of iPSCs in the pig. One of these groups generated porcine iPSCs using fetal fibroblasts from Danish Landrace pigs, by the lentiviral transduction of six human transcription factors (OCT4, SOX2, c-Myc, KLF4, LIN28, and NANOG) under the control of a doxycycline-inducible promoter [41]. Simultaneously, another Chinese group reported the derivation of iPSCs from fibroblasts from the Tibetan minipig, using constitutively expressed lentiviral vectors carrying the mouse cDNA sequences of Oct4, Sox2, c-myc, and Klf4 [42]. The third report published shortly thereafter by Ezashi et al. [43] also described the production of porcine iPSCs through the use of 4 human transcription factors (OCT4, SOX2, c-Myc, and KLF4). In addition, these researchers claimed that the porcine iPSCs were positive for SSEA-1 but negative for SSEA3 and SSEA-4, indicating that they were more similar to mouse than human iPSCs.

Transcriptional profiling of the cell line by Affymetrix microarray confirmed that the cells were indeed reprogrammed, and expressed a variety of ESC markers endogenously. However, the continued expression of the exogenous transcription factors was detected in the iPSCs generated by all three groups. This problem is not unique to the pig, as it has been reported in other species as well [39]. In any case, the continued expression of pluripotency genes did not pose any problems for the differentiation of the cells, since all three laboratories demonstrated that their cell lines were able to differentiate *in vitro* (including EB formation) and in vivo by forming teratomas containing cells of all three germ layers. More recently, it was demonstrated that porcine iPSCs can form chimeras with high efficiency (85.3%), and contribute to all three germ layers [44]. Because germline chimerism was not confirmed in this study, it remains to be determined whether the porcine iPSCs are fully equivalent to mouse iPSCs.

One of the major advantages iPSCs may offer in the future is the potential for custom derivation of pluripotent cells from individual patients to use for regenerative therapies without the risk of immune rejection. However, since the epigenetic reprogramming necessary to produce them requires prolonged expression of the transgenes (2-3 weeks), most of the iPSCs described to date have been generated with the use of lentiviral vectors that integrate known oncogenes (such as c-myc and Klf4) into the cell genome. The dangers of using these genes became evident in a study in which mouse chimeras generated with iPSCs developed tumors following reactivation of the initially silenced transgene c-myc [38]. In the only report to date in which porcine chimeras were produced from iPSCs [44], no tumor formation was reported in the newborn piglets. However, because long-term testing and monitoring were not conducted prior to the publication of these results, at present there is no guarantee that these iPSCs would be safe for clinical applications.

To avoid problems associated with viral integration and the use of oncogenes, many groups have developed strategies for iPSC production in which the disadvantages discussed above have been minimized. For example, it has been shown that reprogramming can be achieved in both human and mouse cells without the use of c-myc, albeit at the expense of efficiency [45]. In addition, the use of certain transcription factors can be omitted when using cell types already expressing them; for example, neural progenitor cells, which already express endogenous Sox2 and c-myc were reprogrammed using only the induced expression of OCT4 and Klf4 [46] [47].

Furthermore, some transcription factors can be replaced by small molecules that have similar effects on somatic cell reprogramming [48]. Viral integration, which carries risks of insertional mutagenesis, can be avoided by using nonintegrating adenoviral vectors [49] [50]. Our laboratory has established porcine iPSCs derived from pig skin fibroblasts using Sendai virus vectors to introduce 4 human transcription factors (OCT4, SOX2, c-Myc, and KLF4) (data not shown). Another attractive alternative to the use of viral vectors are transposon vectors, which combine high transfection efficiency with enhanced safety, and can be removed from the cell genomes following reprogramming. For example, piggyBac is a DNA transposon that, following insertion, can be removed from the reprogrammed genome without leaving a trace. This transposon has already been used successfully to produce iPSCs [51, 52]. Finally, Zhou [53] produced iPSCs by providing the transcription factors in the form of recombinant proteins.

With regard to future research in porcine iPSC production, the choice of cell type and strategy will undoubtedly play an important role in the reprogramming efficiency. It should be noted, however, that the same problems existing in porcine ESC culture are likely to affect the long-term maintenance of the newly generated pig iPSC lines. For example, the culture conditions supporting the pluripotency and self-renewal of the porcine iPSCs apparently require further optimization, given that, in all the published reports to date, the reprogrammed cell lines could be maintained only with the continued expression of exogenous pluripotency genes. In one study, for example, instead of using growth factors or other supplements, the pig iPSCs were maintained with doxycycline-induced expression of the pluripotency transgenes, until the authors chose to differentiate the cells [41].

Despite these challenges, we can without doubt look forward to an exciting future in which iPSCs will add another dimension to pluripotent stem cell research in the pig.

5. Epiblast stem cells

Epiblast stem cells are pluripotent stem cells derived from the epiblast layer of post-implantation mouse embryos [54]. Epiblast stem cells are distinct from ES cells, which are derived from the inner cell mass of blastocysts. Both ES cells and epiblast stem cells can differentiate into mesoderm, endoderm, and ectoderm. However, epiblast stem cells do not have germ-line transmission.

Rodent pluripotent stem cells are considered to have two distinct states: naïve and primed. Naïve pluripotent stem cell lines are distinguished from primed cells in their response to LIF signaling and MEK/GSK3 inhibition (LIF/2i conditions) and X chromosome activation. Human ES and iPS cells both resemble rodent primed epiblast stem cells more closely than rodent naïve ES cells. In addition, iPS cells derived from pigs can obtain the properties of primed epiblast stem cells. iPS cells derived from human, monkey, rabbit, and rat, but not mouse, can also obtain the properties of primed epiblast stem cells [55].

http://dx.doi.org/10.5772/57603

6. Conclusions

Stem cell biology research of the minipig is developing rapidly. Although studies on mouse and human stem cells currently outnumber those of other species, data from these well-studied species provide a good foundation for current and future porcine stem cell research. Despite the challenges associated with developing porcine pluripotent stem cells, recent successes in the fields of induced pluripotent stem cells and somatic stem cells suggest that minipig stem cell research has a promising future.

Acknowledgements

This study was supported in part by a Grant-in-Aid for Young Scientists (B) of the Japan Society for the Promotion of Science (JSPS) and Grant of Strategic Research Foundation Grant-aided Project for Private schools at Heisei 23th from Ministry of Education, Culture, Sports, Science and Technology of Japan, 2011-2015.

Author details

Hideyuki Kobayashi*, Toshihiro Tai, Koichi Nagao and Koichi Nakajima

*Address all correspondence to: hideyukk@med.toho-u.ac.jp

Department of Urology, Toho University School of Medicine, Tokyo, Japan

References

[1] Notarianni E, Laurie S, Moor RM, Evans MJ. Maintenance and differentiation in culture of pluripotential embryonic cell lines from pig blastocysts. J Reprod Fertil Suppl. 1990;41:51-6.

[2] Piedrahita JA, Anderson GB, Bondurant RH. On the isolation of embryonic stem cells: Comparative behavior of murine, porcine and ovine embryos. Theriogenology. 1990;34(5):879-901.

[3] Piedrahita JA, Anderson GB, Bondurant RH. Influence of feeder layer type on the efficiency of isolation of porcine embryo-derived cell lines. Theriogenology. 1990;34(5): 865-77.

[4] Strojek RM, Reed MA, Hoover JL, Wagner TE. A method for cultivating morphologically undifferentiated embryonic stem cells from porcine blastocysts. Theriogenology. 1990;33(4):901-13.

[5] Henderson JK, Draper JS, Baillie HS, Fishel S, Thomson JA, Moore H, et al. Preimplantation human embryos and embryonic stem cells show comparable expression of stage-specific embryonic antigens. Stem Cells. 2002;20(4):329-37.

[6] Hall V. Porcine embryonic stem cells: a possible source for cell replacement therapy. Stem Cell Rev. 2008;4(4):275-82.

[7] Oestrup O, Hall V, Petkov SG, Wolf XA, Hyldig S, Hyttel P. From zygote to implantation: morphological and molecular dynamics during embryo development in the pig. Reprod Domest Anim. 2009;44 Suppl 3:39-49.

[8] Alberio R, Croxall N, Allegrucci C. Pig epiblast stem cells depend on activin/nodal signaling for pluripotency and self-renewal. Stem Cells Dev. 2010;19(10):1627-36.

[9] Reijo Pera RA, DeJonge C, Bossert N, Yao M, Hwa Yang JY, Asadi NB, et al. Gene expression profiles of human inner cell mass cells and embryonic stem cells. Differentiation. 2009;78(1):18-23.

[10] Li M, Ma W, Hou Y, Sun XF, Sun QY, Wang WH. Improved isolation and culture of embryonic stem cells from Chinese miniature pig. J Reprod Dev. 2004;50(2):237-44.

[11] Kim S, Kim JH, Lee E, Jeong YW, Hossein MS, Park SM, et al. Establishment and characterization of embryonic stem-like cells from porcine somatic cell nuclear transfer blastocysts. Zygote. 2010;18(2):93-101.

[12] Chen LR, Shiue YL, Bertolini L, Medrano JF, BonDurant RH, Anderson GB. Establishment of pluripotent cell lines from porcine preimplantation embryos. Theriogenology. 1999;52(2):195-212.

[13] Vassiliev I, Vassilieva S, Beebe LF, Harrison SJ, McIlfatrick SM, Nottle MB. In vitro and in vivo characterization of putative porcine embryonic stem cells. Cell Reprogram. 2010;12(2):223-30.

[14] Greber B, Wu G, Bernemann C, Joo JY, Han DW, Ko K, et al. Conserved and divergent roles of FGF signaling in mouse epiblast stem cells and human embryonic stem cells. Cell Stem Cell. 2010;6(3):215-26.

[15] Hall VJ, Christensen J, Gao Y, Schmidt MH, Hyttel P. Porcine pluripotency cell signaling develops from the inner cell mass to the epiblast during early development. Dev Dyn. 2009;238(8):2014-24.

[16] Matsui Y, Zsebo K, Hogan BL. Derivation of pluripotential embryonic stem cells from murine primordial germ cells in culture. Cell. 1992;70(5):841-7.

[17] Resnick JL, Bixler LS, Cheng L, Donovan PJ. Long-term proliferation of mouse primordial germ cells in culture. Nature. 1992;359(6395):550-1.

[18] Rohwedel J, Sehlmeyer U, Shan J, Meister A, Wobus AM. Primordial germ cell-derived mouse embryonic germ (EG) cells in vitro resemble undifferentiated stem cells with respect to differentiation capacity and cell cycle distribution. Cell Biol Int. 1996;20(8):579-87.

[19] Labosky PA, Barlow DP, Hogan BL. Mouse embryonic germ (EG) cell lines: transmission through the germline and differences in the methylation imprint of insulin-like growth factor 2 receptor (Igf2r) gene compared with embryonic stem (ES) cell lines. Development. 1994;120(11):3197-204.

[20] Pesce M, Farrace MG, Piacentini M, Dolci S, De Felici M. Stem cell factor and leukemia inhibitory factor promote primordial germ cell survival by suppressing programmed cell death (apoptosis). Development. 1993;118(4):1089-94.

[21] Durcova-Hills G, Tang F, Doody G, Tooze R, Surani MA. Reprogramming primordial germ cells into pluripotent stem cells. PLoS One. 2008;3(10):e3531.

[22] Shim H, Gutiérrez-Adán A, Chen LR, BonDurant RH, Behboodi E, Anderson GB. Isolation of pluripotent stem cells from cultured porcine primordial germ cells. Biol Reprod. 1997;57(5):1089-95.

[23] Durcova-Hills G, Prelle K, Müller S, Stojkovic M, Motlik J, Wolf E, et al. Primary culture of porcine PGCs requires LIF and porcine membrane-bound stem cell factor. Zygote. 1998;6(3):271-5.

[24] Mueller S, Prelle K, Rieger N, Petznek H, Lassnig C, Luksch U, et al. Chimeric pigs following blastocyst injection of transgenic porcine primordial germ cells. Mol Reprod Dev. 1999;54(3):244-54.

[25] Lee CK, Piedrahita JA. Effects of growth factors and feeder cells on porcine primordial germ cells in vitro. Cloning. 2000;2(4):197-205.

[26] Lee CK, Weaks RL, Johnson GA, Bazer FW, Piedrahita JA. Effects of protease inhibitors and antioxidants on In vitro survival of porcine primordial germ cells. Biol Reprod. 2000;63(3):887-97.

[27] Petkov SG, Anderson GB. Culture of porcine embryonic germ cells in serum-supplemented and serum-free conditions: the effects of serum and growth factors on primary and long-term culture. Cloning Stem Cells. 2008;10(2):263-76.

[28] Tsung HC, Du ZW, Rui R, Li XL, Bao LP, Wu J, et al. The culture and establishment of embryonic germ (EG) cell lines from Chinese mini swine. Cell Res. 2003;13(3): 195-202.

[29] Petkov SG, Marks H, Klein T, Garcia RS, Gao Y, Stunnenberg H, et al. In vitro culture and characterization of putative porcine embryonic germ cells derived from domestic breeds and Yucatan mini pig embryos at Days 20-24 of gestation. Stem Cell Res. 2011;6(3):226-37.

[30] Turnpenny L, Brickwood S, Spalluto CM, Piper K, Cameron IT, Wilson DI, et al. Derivation of human embryonic germ cells: an alternative source of pluripotent stem cells. Stem Cells. 2003;21(5):598-609.

[31] Shamblott MJ, Axelman J, Littlefield JW, Blumenthal PD, Huggins GR, Cui Y, et al. Human embryonic germ cell derivatives express a broad range of developmentally distinct markers and proliferate extensively in vitro. Proc Natl Acad Sci U S A. 2001;98(1):113-8.

[32] Turnpenny L, Cameron IT, Spalluto CM, Hanley KP, Wilson DI, Hanley NA. Human embryonic germ cells for future neuronal replacement therapy. Brain Res Bull. 2005;68(1-2):76-82.

[33] Hua J, Yu H, Liu S, Dou Z, Sun Y, Jing X, et al. Derivation and characterization of human embryonic germ cells: serum-free culture and differentiation potential. Reprod Biomed Online. 2009;19(2):238-49.

[34] Piedrahita JA, Moore K, Oetama B, Lee CK, Scales N, Ramsoondar J, et al. Generation of transgenic porcine chimeras using primordial germ cell-derived colonies. Biol Reprod. 1998;58(5):1321-9.

[35] Karasiewicz J, Sacharczuk M, Was B, Guszkiewicz A, Korwin-Kossakowski M, Gorniewska M, et al. Experimental embryonic-somatic chimaerism in the sheep confirmed by random amplified polymorphic DNA assay. Int J Dev Biol. 2008;52(2-3): 315-22.

[36] Piliszek A, Modliński JA, Pyśniak K, Karasiewicz J. Foetal fibroblasts introduced to cleaving mouse embryos contribute to full-term development. Reproduction. 2007;133(1):207-18.

[37] Takahashi K, Yamanaka S. Induction of pluripotent stem cells from mouse embryonic and adult fibroblast cultures by defined factors. Cell. 2006;126(4):663-76.

[38] Okita K, Ichisaka T, Yamanaka S. Generation of germline-competent induced pluripotent stem cells. Nature. 2007;448(7151):313-7.

[39] Takahashi K, Tanabe K, Ohnuki M, Narita M, Ichisaka T, Tomoda K, et al. Induction of pluripotent stem cells from adult human fibroblasts by defined factors. Cell. 2007;131(5):861-72.

[40] Yu J, Vodyanik MA, Smuga-Otto K, Antosiewicz-Bourget J, Frane JL, Tian S, et al. Induced pluripotent stem cell lines derived from human somatic cells. Science. 2007;318(5858):1917-20.

[41] Wu Z, Chen J, Ren J, Bao L, Liao J, Cui C, et al. Generation of pig induced pluripotent stem cells with a drug-inducible system. J Mol Cell Biol. 2009;1(1):46-54.

[42] Esteban MA, Xu J, Yang J, Peng M, Qin D, Li W, et al. Generation of induced pluripotent stem cell lines from Tibetan miniature pig. The Journal of biological chemistry. 2009;284(26):17634-40.

[43] Ezashi T, Telugu BP, Alexenko AP, Sachdev S, Sinha S, Roberts RM. Derivation of induced pluripotent stem cells from pig somatic cells. Proc Natl Acad Sci U S A. 2009;106(27):10993-8.

[44] West FD, Terlouw SL, Kwon DJ, Mumaw JL, Dhara SK, Hasneen K, et al. Porcine induced pluripotent stem cells produce chimeric offspring. Stem Cells Dev. 2010;19(8): 1211-20.

[45] Nakagawa M, Koyanagi M, Tanabe K, Takahashi K, Ichisaka T, Aoi T, et al. Generation of induced pluripotent stem cells without Myc from mouse and human fibroblasts. Nat Biotechnol. 2008;26(1):101-6.

[46] Kim JB, Zaehres H, Wu G, Gentile L, Ko K, Sebastiano V, et al. Pluripotent stem cells induced from adult neural stem cells by reprogramming with two factors. Nature. 2008;454(7204):646-50.

[47] Shi Y, Desponts C, Do JT, Hahm HS, Schöler HR, Ding S. Induction of pluripotent stem cells from mouse embryonic fibroblasts by Oct4 and Klf4 with small-molecule compounds. Cell Stem Cell. 2008;3(5):568-74.

[48] Lin T, Ambasudhan R, Yuan X, Li W, Hilcove S, Abujarour R, et al. A chemical platform for improved induction of human iPSCs. Nat Methods. 2009;6(11):805-8.

[49] Okita K, Nakagawa M, Hyenjong H, Ichisaka T, Yamanaka S. Generation of mouse induced pluripotent stem cells without viral vectors. Science. 2008;322(5903):949-53.

[50] Stadtfeld M, Nagaya M, Utikal J, Weir G, Hochedlinger K. Induced pluripotent stem cells generated without viral integration. Science. 2008;322(5903):945-9.

[51] Woltjen K, Michael IP, Mohseni P, Desai R, Mileikovsky M, Hämäläinen R, et al. piggyBac transposition reprograms fibroblasts to induced pluripotent stem cells. Nature. 2009;458(7239):766-70.

[52] Yusa K, Rad R, Takeda J, Bradley A. Generation of transgene-free induced pluripotent mouse stem cells by the piggyBac transposon. Nat Methods. 2009;6(5):363-9.

[53] Zhou H, Wu S, Joo JY, Zhu S, Han DW, Lin T, et al. Generation of induced pluripotent stem cells using recombinant proteins. Cell Stem Cell. 2009;4(5):381-4.

[54] Brons IG, Smithers LE, Trotter MW, Rugg-Gunn P, Sun B, Chuva de Sousa Lopes SM, et al. Derivation of pluripotent epiblast stem cells from mammalian embryos. Nature. 2007;448(7150):191-5.

[55] De Los Angeles A, Loh YH, Tesar PJ, Daley GQ. Accessing naive human pluripotency. Current opinion in genetics & development. 2012;22(3):272-82.

Section 5

Ethics

Chapter 10

Ethical Implications of Embryo Adoption

Peter A. Clark

Additional information is available at the end of the chapter

http://dx.doi.org/10.5772/58592

1. Introduction

It is estimated that 2.1 million married couples or 5 million people in the United States are affected by infertility.[1] Infertility is defined as failure to get pregnant after one year of unprotected intercourse. About 40% of infertility cases are due to a female factor and 40% due to a male factor. The remaining 20% are the result of a combination of male and female factors, or are of unknown causes. [2] Issues of human infertility are extremely complex physiologically, psychologically, financially, legally and ethically. It is estimated that 85-90% of infertile couples will receive conventional treatment and 10-15% may become candidates for various forms of Assisted Reproductive Technologies (ARTs) to assist them in having their own biological children. In-vitro fertilization (IVF) is one of the most utilized reproductive procedures that has allowed couples to have their own biological children. IVF accounts for 99% of ART. This procedure has been effective but it is still inefficient and expensive. One aspect of the inefficiency is that numerous embryos have been frozen through a process called cryopreservation. It has been estimated that there are 400,000 embryos frozen and stored since the late 1970s. [3] In reality, the actual number of frozen embryos is probably closer to 500,000 with an additional 20,000 embryos added yearly. [4] Freezing these embryos has allowed for a limitation on the number of embryos transferred to a woman's uterus which has decreased the number of multiple gestations. It also allows couples to use the frozen embryos in the future if the initial cycles are unsuccessful. This is not only more effective but also lowers the cost. The issue is now what to do with the 400,000 to 500,000 frozen embryos that remain as "spares." Various alternatives have been suggested. The embryos could be thawed and then destroyed, continued to be cryopreserved indefinitely, used for research, or offered for donation/adoption. All of these options present problems medically, legally and ethically.

Medically, the lifespan of a cryopreserved embryo is unknown. The effect of the freezing process is also unknown on the quality of the embryo if brought to term. "Studies have found that babies created through IVF are twice as likely to be born underweight and with major birth defects." [5] With the unknown effects of cryopreservation on embryo development the medical issues become even more complex. Legally, only 2% of frozen embryos are specifically designated for donation/adoption and 5% are specifically designated for destruction or research. [6] The legal issues focus on the applicability of contract law versus family law because frozen embryos are technically considered "property" not "persons." Presently, the applicability of contract law or family law remains unclear. In addition, to date only three states —Florida, Louisiana and New Hampshire—have adopted legislation concerning the disposition or disposal of embryos. Legally and legislatively the issue of embryo donation/adoption is ambiguous at best. Ethically, depending on one's view of when personhood begins, frozen embryos may be considered human persons, which deserve dignity and respect, or they may have less than human status with no particular ethical rights. From an ethical perspective that views personhood beginning at fertilization, one could argue that the "rescue" of these embryos would not only be ethically acceptable but morally mandatory. To determine if frozen embryos should be donated/adopted all of these issues will have to be examined.

This article will focus on embryo donation/adoption as a viable option to address the 400,000 to 500,000 frozen embryos in the United States. The intended purpose of this article is fourfold: first, to examine the medical issues surrounding the cryopreservation of frozen embryos; second, to examine the legal issues that focus on the applicability of contract law and family law; third, to give an ethical analysis of the arguments for and against embryo donation/adoption; and fourth, to give recommendations on how to avoid the continuation of this problem in the future.

2. Medical aspects

Infertility is a major problem for many couples in the United States. "About one married couple in 12 cannot conceive a child after two years of trying. Infertility stems from many factors, including a woman's age at the first attempt to conceive, damage from pelvic inflammatory disease, previous abortions, uterine abnormalities, and a man's low sperm count or low sperm motility." [7] Individually, male and female factors each account for about 40% of infertility in the United States. Numerous technologies are available to couples from artificial insemination by a husband or a donor, to gamete intrafallopian transfer (GIFT), to zygote intrafallopian transfer (ZIFT), to in-vitro fertilization. Of these reproductive technologies IVF has become the ART of choice for many infertile couples. IVF is an assisted reproductive technology which had its first success in 1978 when Drs. Edwards and Steptoe in Oldham, England created the first "test tube baby" named Louise Brown. Since that first success, IVF technology has been refined and over 3 million babies have been born worldwide. [8]

There are five basic steps to IVF. *1) Harvesting the eggs from the woman's ovaries.* The woman's ovaries are hyperstimulated using fertility drugs that produce numerous eggs. During this period the woman will have regular transvaginal ultrasounds to examine the ovaries and blood tests to check hormone levels. *2) Egg retrieval.* The eggs are removed from the woman's body using follicular aspiration. Using ultrasound images as a guide the physician inserts a thin

http://dx.doi.org/10.5772/58592

needle through the vagina and into the ovary and sacs containing the eggs. The needle is connected to a suction device, which pulls the eggs and fluid out of each follicle, one at a time. In rare cases, a pelvic laparoscopy may be used to remove the eggs. *3) Insemination and Fertilization.* The man's sperm is placed with the best quality eggs in a petri dish and stored in an environmentally controlled chamber. The mixing of the sperm and egg is called insemination. The sperm usually enters an egg a few hours after insemination. If there is a low chance for fertilization, one single sperm can be injected into an egg in a procedure called Intracytoplasmic Sperm Injection (ICSI). *4) Embryo culture.* The fertilized eggs remain in the petri dish for 48 to 72 hours to verify that the embryo is not defective and growing properly. If a couple is at high-risk for passing on genetic (hereditary) disorders to a child they may consider using Pre-implantation Genetic Diagnosis (PGD). The procedure is performed 3-4 days after fertilization. A single cell is removed from each embryo to screen it for specific genetic disorders. Those embryos with the genetic disorder are usually destroyed. *5) Embryo transfer.* Anywhere from 1-4 embryos are placed in the woman's womb 3 to 4 days after fertilization. The physician inserts a thin catheter containing the embryos into the woman's vagina, through the cervix, and up into the womb. If the embryo implants in the woman's uterine wall pregnancy will result. [9]

The implantation rate is estimated at 10-25%. [10] The overall birth rate varies from 11% (women over 40) to about 35% (women under 35). [11] This clearly shows that a number of embryos transferred fail to survive, which is why multiple embryos are transferred per cycle and why numerous cycles are required. On average, 2.7 embryos per cycle are transferred in women under 35, with an average of 3 in older women. Depending on the embryo quality, up to 5-6 embryos can be transferred. [12] The average cost of IVF is $12,000-17,000 per cycle. It is estimated that 75% of couples who have tried IVF and who spent from $10,000-100,000 still go home without a baby. [13] Risks include the possibility of ovarian hyperstimulation syndrome (OHSS), risks in the egg retrieval stage which include reactions to anesthesia, bleeding, infection and damage to structures surrounding the ovaries including the bowel and bladder, and finally there are the risks associated with multiple pregnancies. Since 1980 the rate of twins has climbed 70% to 3.2% of births in 2004. Multiple gestations raise the risk of preterm births; low-birth-weight babies, with the possibility of death in very premature infants; long-term health problems; and pregnancy complications, which include pre-eclampsia, gestational diabetes, and Caesarean section. Studies have shown that 56% of IVF twins born in 2004 weighed less than 5.5 pounds, and 65% were born prematurely, before 37 weeks of gestation. [14] Embryos not transferred in a fresh IVF cycle are usually cryopreserved. Freezing these embryos offers individuals the possibility of transferring the frozen embryos for later IVF cycles if the previous cycle does not result in a pregnancy. It is also cost effective and eliminates the need to undergo the steps needed for a fresh IVF cycle. In most cases the best quality embryos are transferred in the fresh cycle and those of a lesser quality are frozen for later transfer. It should be noted that some clinics have individual freezing and thawing to achieve the exact number of embryos desired for transfer. This procedure avoids embryo wastage.

The process of cryopreservation has become an integral part of the IVF procedure. "Cryopreservation is a process of freezing biological tissues for storage, while minimizing cellular damage from freezing and thawing." [15] This technique entails freezing the embryo while simultaneously removing the intracellular water and replacing it with a cryoprotectant solution which help to protect the embryo during the freezing process. The embryos are then placed into cryopreservation straws or vials, which are labeled with the patient's name, the patient's IVF number, and the date of the freeze. Once the process is complete, the embryos are placed in a computer controlled freezing unit. After the freezing run is complete, the straws are stored in a special tank filled with liquid nitrogen at a temperature of minus 196 degrees centigrade. [16] Many storage facilities use a back-up system to minimize the risk of interruption in the freezing process. Liquid nitrogen containers are armed with an automatic alarm system to monitor nitrogen levels and prevent premature thawing. [17] These embryos are looked upon as being in a state of "suspended animation." Cellular activity has ceased, but each embryo is still alive. When the remaining embryos are needed a procedure utilizing rapid thawing and removal of the cryopreservative solution with simultaneous rehydration is used. The embryos are first warmed in a 98.6 F degree solution and the cryoprotectant chemicals are removed. [18]

The embryo thawing process is quite complex. "Embryo survival is based on the number of viable cells in an embryo after thawing. An embryo has 'survived' if >50% of the cells are viable. An embryo is considered to 'partially survive' if <50% of its cells are viable and to be 'atretic' if all the cells are dead at thaw. Approximately, 65-70% of embryos survive thaw, 10% partially survive and 20-25% are atretic. Data suggests that embryos with 100% cell survival are almost as good as embryos never frozen but only about 30-35% survive this fashion. Embryos that are 2, 4 or 8 cells when frozen have about a 5-10% greater survival than embryos with an odd number of cells. Donor egg embryos have a 2-5% greater survival rate than embryos from infertile women when compared by morphology score" [19] The cost of cryopreservation is approximately $600-700 a year. The success rate or pregnancy rate depends on numerous factors: the number of surviving embryos transferred, the number of 100% surviving embryos transferred, and the morphology scores of the transferred embryos. The delivered pregnancy rates range from 5% (a single poor quality embryo) to 36% (4 high quality embryos) when the cycles from 1987 to 2001 were combined. It is estimated that embryo cryopreservation adds about 10-30% more pregnancies per retrieval cycle and the outcomes of the children are normal. [20] The reason for the wide range of costs and success rates is because the Assisted Reproductive Technologies industry in the United States is unregulated. The success rates and costs can vary from clinic to clinic and there is no government oversight examining the widespread differences.

The RAND/SART survey in 2003 found that of the 400,000 frozen spare embryos 88.2% were designated for family building and 2.8% (11,000) were designated for research. Those embryos designated for research could produce as many as 275 stem cell lines (cell cultures suitable for further development). However, the number would in reality be much lower. Of the remaining embryos, it is estimated that 2.3% (10,000) are awaiting donation, 2.2% are designated to be discarded, and 4.5% are held in storage for other reasons, including lost contact with a patient,

http://dx.doi.org/10.5772/58592

patient death, abandonment, and divorce. [21] There are numerous issues concerning the "spare" frozen embryos. The ART clinics transfer the highest quality embryos (those that grow at a normal rate) to the patient during treatment cycles. The remaining embryos are usually designated as not of the highest quality. In addition, some of the frozen embryos have been in storage for many years, and when these embryos were created the laboratory cultures were not as conducive to preserving embryos as they are today. Some embryos would also die in the freeze-thaw process. Considering all these issues, the question is how many embryos actually are available for research and donation/adoption? The RAND/SART team estimated that 65% of the approximately 11,000 embryos designated for research would survive the freeze-thaw process, resulting in 7,334 embryos. Of those, about 25% (1,834 embryos) would likely be able to survive the initial stages of development to the blastocyst stage (a balstocyst is an embryo that has developed for at least 5 days). Even fewer could be converted into embryonic stem cell lines. Their estimate is about 275 embryonic stem cell lines could be converted from the total number of embryos designated for research. The RAND/SART team also estimates that 2.3% of the 400,000 frozen "spare" embryos designated for donation/adoption, only 23,000-100,000 embryos could be adopted, thawed and successfully born. [22] Having this many children potentially available for adoption would help meet the need of couples seeking adoption in the United States. The problem is that the adoption process for frozen embryos is quite ambiguous and very complex.

3. Legal aspects

There are approximately 200,000 couples actively seeking to adopt in the United States. Having the potential of 23,000-100,000 embryos available to be adopted, thawed and successfully born would offer great hope to these couples. Organizations like Nightlight Christian Adoptions, licensed in California since 1959, arrange both domestic and international adoptions. Their Snowflake Embryo Adoption Program, which began in 1997, matches couples who have spare frozen embryos with other infertile couples trying to have babies. Their philosophy is that every embryo is a person from the minute it exists in a petri dish. Nightlight Christian Adoptions approached embryo adoption differently from other agencies. "Snowflake goes beyond the embryo donation provided by fertility clinics by offering safeguards and education available in traditional adoption. A home study is prepared on the adopting family that includes screening and education. The donating family is responsible for selecting a family to raise their genetic child (as opposed to a doctor in a clinic making the selection for the family), and they will know if the child (children) is born from the adopted embryos. Our program recognizes the importance of counseling all parties involved. Most importantly, at Nightlight we recognize the personhood of embryos and we treat them as precious preborn children." [23] There are no agency or program fees for the genetic parents who place their embryos for adoption. Any costs during the adoption process for medical records, blood work, etc., will be paid by the adopting parents. Fees differ for in-state California residents and out-of-state residents. If you live outside of California the Program Fee is $8000; fee for the agency performing the home study ranges from $1000-3000; and the Fertility Clinic's Fee for a Frozen Embryo Transfer (FET) ranges from $2000-7500. In-state residents pay a Program Fee of $10,600. A $2600 credit is applied if you already completed a home study with another agency. The Fertility Clinic's Fee for FET ranges from $2000-7500. [24] By contrast, the National Embryo Donation Center estimates the cost of embryo adoption to be $4,560-5,360. That includes the Application Fee $200 (international application fee is $300); Program Fee (to proceed to assessment for embryo transfer) $800; Embryo Transfer $650; Embryology Laboratory Fee $565; Monitoring Fee $250; Facility Fee $700; Home Study $1000-2000; Initial Consult Fee $200; and Trial Transfer Fee $85. The National Average for IVF is $7500-9000/ cycle and the National Average for IVF with Donor Egg is $22,127. [25] It is clear that the price differential is considerable. Recent statistics show that Snowflake has matched 289 placing families (with approximately 2,092 embryos) with 192 adopting families. 139 babies have been born and 14 adopting families are currently expecting 15 babies. [26]

The legal issues focus on the terminology surrounding adoption and donation. The term "adoption" raises opposition with abortion-rights groups because it encourages people to view the frozen "spare" embryos as equivalent to children. These groups would prefer the term "embryo donation," or in more neutral, reductive terms, a term such as "transfer of genetic material" from one party to another. [27] The distinction between "embryo adoption" and "embryo donation" may seem trivial to many but from a legal perspective it raises numerous issues. The Supreme Court of Tennessee in *Davis v. Davis* recognized that, "semantical distinctions are significant in this context because language defines legal status and can limit legal rights." [28] The court in *Davis v.Davis* also concluded that pre-embryos are not, strictly speaking, either persons or property, but occupy an interim category that entitles them to special respect because of their potential for human life. [29] The American Society of Reproductive Medicine has echoed this conclusion: "The embryo deserves respect greater than that accorded to human tissue but not the respect accorded to actual persons. The embryo is due greater respect than human tissue because of its potential to become a person and because of its symbolic meaning for many people. Yet, it should not be treated as a person, because it has not yet developed the features of personhood, is not yet established as developmentally individual, and may never realize its biological potential." [30] The conclusion seems to indicate that neither contract law nor family law can directly interpret embryo donation/adoption agreements. Contract law governs the transfer of property, while family law governs lives of persons in familial relationships. If embryos are neither property nor persons, but an interim category, it follows that a hybrid approach must be considered. [31]

Parties involved with embryo donation/adoption need certainty concerning their contractual rights and obligations. "Unlike traditional adoption, which has multiple procedural requirements, embryo donation is largely unregulated. Some commentators warn that calling an embryo donation an "embryo adoption" may give the recipient parents a false sense of security regarding their parental rights and responsibilities since most states do not extend traditional adoption laws to the adoption of an embryo. Additionally, both state laws and the Uniform Adoption Act consistently state that children cannot be adopted until after they are born." [32]

Because the law is so ambiguous on this topic it would appear that the state legislatures or the federal government would be the appropriate forum to address these issues. As one court noted:

> We must call on the Legislature to sort out the parental rights and responsibilities of those involved in artificial reproduction. No matter what one thinks of artificial insemination, traditional and gestational surrogacy (in all its permutations), and—as now appears in the not-too-distant future, cloning and even gene splicing—courts are still going to be faced with the problem of determining lawful parentage. A child cannot be ignored. [33]

A few states have begun to enact legislation regarding embryo donation/adoption, but in reality most states lack appropriate statutes. In Florida, a donated embryo is presumed to be a child of the intended parents if both the donor couple and the intended parents consent in writing. The statute effectively requires the donor couple to relinquish their parental rights, but the statute does not specify how this is to be accomplished. [34] In Oklahoma the statute requires that both the donor and the intended parents must be married and the physician performing the transfer must obtain written consent from both the donor and the intended parents. This consent form must be signed by both the physician and the judge of a court with adoption jurisdiction. The original consent form is then filed with the court by the physician. Any child resulting from the embryo donation is considered to be the child of the donee couple and the donee couple is relieved of all parental responsibilities. [35] Worldwide embryo adoption is performed in at least 19 countries (Canada, UK, France, Spain, Italy, Australia, Belgium, India, Greece, Singapore, Argentine, Colombia, Japan, Holland, Uruguay, Romania, Portugal, Venezuela and Finland). Embryo Adoption is illegal in 14 countries (Austria, China, Denmark, Germany, Israel, Italy, Latvia, Norway, Slovenia, Sweden, Switzerland, Taiwan, Tunisia and Turkey). In the United States all 50 states and the District of Columbia permit living embryo adoption and implantation. [36] The problem is that there is real uncertainty in the law and some might even say it is chaotic. It appears that legislation is needed to protect the rights of these embryos, their biological parents and their adopted parents. Issues concerning legislation range from disagreement about whether this legislation should be initiated from the states or from the federal government to ambiguities concerning personhood and how this will impact on current legal statutes. Legislation appears to be the only route available to overcome the ambiguity in the law. However, legislators are looking for guidance and one area that might offer such assistance is the realm of ethics.

4. Ethical aspects

Ethically, embryo donation/adoption focuses on the issue of personhood. If embryos are persons then it would be a moral imperative to "rescue" these embryos from their current status of being in "frozen animation." Numerous ethicists, embryologists, legal professionals and specifically, the Roman Catholic Church, argue that personhood begins at conception or what is known as fertilization. Prior to fertilization we have two human gametes—sperm and egg, that are living but are not a living organism. When fertilization occurs, something human and living "in a different sense comes into being." [37] Embryologists argue that "human development begins at fertilization when a male gamete or sperm (spermatozoon) unites with a female gamete or oocyte (ovum) to form a single cell—zygote. This highly specialized, totipotent cell marked the beginning of each of us as a unique individual." [38] The Catholic Church teaches that "human life must be absolutely respected and protected from the moment of conception." [39] "Right from fertilization is begun the adventure of a human life, and each of its great capacities requires time...to find its place and to be in a position to act. This teaching remains valid and is further confirmed, if confirmation were needed, by recent findings of human biological science which recognize that in the zygote resulting from fertilization the biological identity of a new human individual is already constituted." [40] The Church argues that at fertilization there is a new genetic individual in its own right, one who is whole, bodily, self-organizing, and genetically distinct from his or her mother and father. [41] Those who argue that personhood begins at fertilization would also argue that there is a moral imperative to give these frozen embryos the opportunity to be born and to develop because they are persons. Ethicist Therese Lysaught believes that embryo donation/adoption is an act that can properly be described as "rescuing a child orphaned before birth." [42] Ethicists arguing for the "rescue" of these children would encourage women to implant these embryos in their wombs in order to bring them to term. Some would permit not only married women to do this but also single women and even lesbian couples. The moral principle of sanctity of human life would overcome any other moral considerations. However, not all, even in the Catholic Church, would agree to this ethical analysis. Opponents of this position argue that this would amount to material cooperation in an objective immoral action. Not only is the process of IVF considered an intrinsic moral evil by the Magisterium of the Catholic Church, but allowing for the adoption of these embryos might condone the objective immoral procedure and may even encourage the creation of additional embryos through the IVF process. Even though the Catholic Church has not taken an official position on embryo donation/adoption, one could argue that from previous teaching, it is the only means of survival for these persons. "In consequence of the fact that they have been produced in vitro, those embryos which are not transferred into the body of the mother and are called 'spares' are exposed to an absurd fate, with no possibility of their being offered safe means of survival which can be licitly pursued." [43] Embryo donation/adoption is the only safe means of survival for these persons so thus it would be ethical. This statement by the Magisterium was directed toward embryo experimentation but it could also be applicable to embryo donation/adoption. To determine if embryo donation/adoption is ethical and to address the ambiguities and unresolved issues surrounding this controversy, the traditional ethical principle of the lesser of two evils will be applied to this situation.

Society, in general, has always recognized that in our complex world there is the possibility that we may be faced with conflict situations that leave us with two options both of which are nonmoral evils. [44] The time-honored ethical principle that has been applied to these situations is called the principle of the lesser of the two evils. When one is faced with two options, both of which involve unavoidable (nonmoral) evil, one ought to choose the lesser evil. [45] Bioethicist Richard McCormick, S.J., argues that

---

The concomitant of either course of action is harm of some sort. Now in situations of this kind, the rule of Christian reason, if we are governed by the *ordo bonorum*, is to choose the lesser evil. This general statement is, it would seem, beyond debate; for the only alternative is that in conflict situations we should choose the greater evil, which is patently absurd. This means that all concrete rules and distinctions are subsidiary to this and hence valid to the extent that they actually convey to us what is factually the lesser evil... Now, if in a conflict situation one does what is, in balanced Christian judgment (and in this sense objectively), the lesser evil, his intentionality must be said to be integral. It is in this larger sense that I would attempt to read Thomas Aquinas's statement that moral acts *recipiunt speciem secundum id quod intenditur*. Thus the basic category for conflict situations is the lesser evil, or avoidable/unavoidable evil, or proportionate reason. [46]

Therefore, in a conflict situation, an individual may directly choose to do a nonmoral evil (violating the person's autonomy, privacy, etc.) as a means to a truly proportionate good end (preservation and protection of human life). [47]

The principle of the lesser of two evils is applicable to the issue of embryo donation/adoption because one is faced with two options, both of which involve unavoidable nonmoral evils. On the one hand, failure to thaw, transfer and allow these embryos to be born would result in the death of thousands of persons. On the other hand, if the frozen embryos are not donated/ adopted they will be discarded, destroyed for research purposes, abandoned, or left in "suspended animation" indefinitely, which would continue to jeopardize their life.

The direct intention of embryo donation/adoption is to protect and preserve human life by saving the lives of vulnerable at-risk embryos. It would also lessen significant hardship associated with ova harvesting, reduce the cost of infertility treatments, and would overcome the objections of couples who resist traditional adoption by allowing the mothers to bond with the child in pregnancy. [48] However, in the process of protecting and preserving human life and acting in the best interest of the frozen embryo, the autonomy of parents might be violated in that some may wish to discard the embryos, allow them to be destroyed to obtain embryonic stem cells, abandon them or allow them to stay in indefinite "suspended animation." The hope is that couples would voluntarily agree to embryo donation/adoption, but studies have shown that only 2% of couples with frozen embryos wish to allow them to be donated or adopted. About 5% are designated for destruction or research which leaves about 87% that are undecided about disposition of their remaining frozen embryos. [49] The linchpin for resolving which option is the lesser of two evils rests on whether or not there is a proportionate reason for allowing embryo donation/adoption.

Proportionate reason refers to a specific value and its relation to all elements (including nonmoral evils) in the action. [50] The specific value in allowing for embryo donation/adoption is to protect and preserve human life. The nonmoral evil, which is the result of trying to achieve this value, is the violation of the couple's right to privacy and autonomy to allow the frozen embryos to be discarded, destroyed for research, abandoned, or left in "suspended animation" indefinitely. The ethical question is whether the value of protecting and preserving human life outweighs the nonmoral evil of violating a couple's right to privacy and autonomy? To determine if a proper relationship exists between the specific value and the other elements of the act, ethicist Richard McCormick, S.J. proposes three criteria for the establishment of proportionate reason:

1. The means used will not cause more harm than necessary to achieve the value.

2. No less harmful way exists to protect the value.

3. The means used to achieve the value will not undermine it. [51]

The application of McCormick's criteria to embryo donation/adoption supports the argument that there is a proportionate reason for allowing these embryos to be thawed, transferred and brought to term. The bottom line is that these embryos already exist and therefore, the preservation of their lives takes moral precedence over any other consideration. First, it is estimated that the average couple who undergoes IVF has seven embryos in storage; the average storage period is four years; and 87% of IVF couples are 'undecided' as to the disposition of their remaining frozen embryos. It is estimated that 23,000 to 100,000 children could be adopted, thawed and successfully born from the 400,000 to 500,000 live human embryos stored at present. [52] Some opponents argue that these embryos are vital to embryonic stem cell research. Allowing for donation/adoption will have an adverse effect on our embryonic stem cell research program. The RAND/SART researches calculated that about 275 embryonic stem cell lines could be created from the total number of embryos available for research. However, they argue that even this number is probably an overestimate because it assumes that all the embryos designated for research in the United States would be used to create stem cell lines, which is highly unlikely. [53] Considering the new methods being proposed to obtain embryonic stem cells such as modified therapeutic cloning, reprogramming of skin cells to their embryonic stage, etc., and the condition of the frozen embryos after thawing, it appears that using these frozen embryos for research purposes would not be in the best interest of the scientific community. There are approximately 200,000 couples seeking to adopt children in the United States. The cost of infertility treatments place ART out of reach for many of these couples. Traditional adoption is also quite expensive and denies couples the chance to experience pregnancy, bonding and breastfreeding that makes the experience "theirs." Embryo donation/adoption allows couples or single women to preserve the lives of already existing embryos which is acting in their best interest. This means gestation by a couple or a single woman who will assume full parental authority for the child. Clearly, this will bring about more good than harm, and will cause less harm than necessary to protect and save lives.

Second, at present, there does not appear to be an alternative that is as effective as embryo donation/adoption to protect and preserve the value of the human lives that are presently in "suspended animation." There are three alternatives to embryo donation/adoption: discarding of the embryos, destruction of the embryos for research purposes and allowing the embryos to stay in "suspended animation" indefinitely. None of these alternatives will protect and preserve the value of the life of the embryo. There is a concern that the length of time embryos are kept in frozen storage may have a detrimental effect on the outcome of embryo transfer and possibly increase fetal abnormalities. To date, no long-term studies have been carried out since the age of the oldest child born as a result of frozen embryo transfer 14 years ago. [54] In addition, according the Genetics and IVF Institute, "Approximately 65-70% of embryos survive thaw, 10% partially survive and 20-25% are atretic." [55] Subjecting embryos to the freeze-thaw process is placing them at significant risk of harm and possibly death. Intentionally or unintentionally, frozen embryos have the potential to be damaged and destroyed. Being in the category of having a special status, embryos deserve not to be harmed or killed. Embryo donation/adoption is the only alternative that protects and preserves the life of the already existing embryo. In the United States there seems to be a consensus that these embryos deserve special respect. This led the Ethics Committee of the American Fertility Society to conclude:

> We find a widespread consensus that the pre-embryo is not a person but is to be treated with special respect because it is a genetically unique, living human entity that might become a person. In cases in which the transfer to a uterus is possible, special respect is necessary to protect the welfare of the potential offspring. In that case, the pre-embryo deserves respect because it might come into existence as a person. This viewpoint imposes the traditional duty of reasonable prenatal care when actions risk harm to prospective offspring. Research on or intervention with a pre-embryo, followed by transfer, thus creates obligations not to hurt or injure the offspring who might be born after transfer. [56]

Whether one believes the frozen embryo is a person or a potential person, it seems clear that this human entity deserves dignity and respect. The only option that would allow for this dignity and respect is to allow for the protection and preservation of the human embryo through embryo donation/adoption.

Third, embryo donation/adoption does not undermine the value of human life. One can argue convincingly that the intention of embryo donation/adoption is to protect and preserve the lives of already existing embryos that are currently in the state of "suspended animation." Those who adopt these embryos have the best interest of the embryos as their primary concern, because they wish to allow the embryos to resume their natural development and growth. The couples and individuals who bring these embryos to term are also willing to adopt these children and take full responsibility for their upbringing in the future. In many situations, couples allow for cryopreservation of embryos because it saves both time and money in the event that the previous cycle of IVF is unsuccessful. This undermines the basic value of human life, because it commodifies, objectifies and exploits these embryos. Allowing the frozen embryos to be discarded, destroyed for research purposes, abandoned or left in the state of "suspended animation" undermines the value of human life. The only possible consequence of this action is the potential destruction of human life.

The intention of embryo donation/adoption is to save lives and it has been proven through organizations such as the National Embryo Donation Center and Nightlight Christian Adoptions to be effective. This is a critical issue that must be addressed immediately because innocent lives are hanging in the balance. It seems clear that there is a proportionate reason for allowing embryo donation/adoption. It is estimated that 23,000-100,000 children could potentially be born as a result of embryo donation/adoption. Couples who are unable to afford ART would have a viable option of having a child that is within their financial means. Finally, safeguards could be put in place that would eliminate creating "spare" embryos in the future. Therefore, it is ethically justified under the principle of proportionate reason for allowing embryo donation/adoption. Embryo donation/adoption is the lesser of two evils because the greater good is promoted in spite of the potential for evil consequences.

5. Conclusion & safeguards

Embryo donation/adoption is a complex issue that has medical, legal and ethical dimensions. Allowing for embryo donation/adoption is the only viable option that protects and preserves their human life. The other viable options: being discarded, destroyed for research, abandoned or kept in "suspended animation" indefinitely, are unacceptable because they have the potential of harming or intentionally killing these embryos that deserve special respect.

To make sure that this situation does not continue in the future, the following recommendations and safeguards are proposed:

1. Only the number of eggs to be placed in the uterus of the mother will be fertilized. Embryos must not be subjected to an intentional interruption of their natural growth and development. There will no longer be "spare" embryos subjected to cryopreservation. Only cryopreservation of gametes would be acceptable.

2. Nationally, laws and legislation must be enacted at the federal level that begins to regulate Assisted Reproductive Technologies. Having each state governed by differing sets of legislation could cause potential complications associated with the practice of donation/adoption. How each state defines jurisdiction and how each state interprets at what stage jurisdiction would begin (conception, transfer, or birth) could become highly complex. Specifically, guidelines and safeguards must be put in place that protects donors, parents, providers, and children born of ART.

3. Nationally, laws and legislation must be enacted that regulates the creation, destruction and exploitation of human embryos. Example would be the following: a) legislation established in New Mexico stating that human embryos can only be disposed of through

implantation, not intentional destruction or through destructive human embryo research. b) Embryos must not be subjected to non-therapeutic experimentation.

4. Internationally, due to the globalization of medical research, it would be imperative for the United Nations (UN) of the World Health Organization (WHO) to establish a forum to study these issues and to formulate a comprehensive policy to regulate the creation of spare embryos and other pertinent issues related to this issue. Both the UN and the WHO have agencies that could bring about legislative mandates.

5. Infertile couples and individuals willing to take full responsibility for the upbringing of these children should be encouraged to consider adoption of the presently existing frozen embryos.

6. Children who are adopted from frozen embryos have the right to know their genetic make-up. They should be given full access to documentation about their biological mothers and fathers so that if this information is needed in the future it is available. This does not mean they have the right to know the names of their biological parents. The right of privacy of the biological parents should be respected.

If we as a nation truly believe that human life deserves dignity and respect, then our failure to bring these embryos to term would be medically irresponsible and ethically objectionable.

Author details

Peter A. Clark*

Address all correspondence to: pclark@sju.edu

Institute of Catholic Bioethics, Saint Joseph's University, Philadelphia, USA

References

[1] National Center for Health Statistics, "Infertility," October 31, 2007. http://www.cdc.gov/nchs/fastats/fertile.htm.

[2] Center for Disease Control and Prevention, "Assisted Reproductive Technology," June 2008. Http://www.cdc.gov/ART/

[3] Rand Law and Health/Society of Assisted Reproductive Technology (SART) Working Group, "How Many Frozen Human Embryos Are Available For Research?" May 2003. http://www.rand.org/pubs/research_briefs/RB9038/index1.html.

[4] Papanikolaou EG, Camus M, Kolibianakis EM, Van Landuyt L, Van Steirteghem A, Devroey P. "In Vitro Fertilization with Single Blastocyst-Stage versus Single Cleavage-Stage Embryos." *New England Journal of Medicine* 354 (March 16, 2006): 1139-1146.

[5] Lysaught, Therese. "Embryo Adoption?" *Human Life Review* (2003): 1-4.. http://findarticles.com/p/articles/mi_qa3798/is_200310/ai_n9297175/print.

[6] Collins, Timothy. "On Abandoned Embryos," *The Linacre Quarterly* 75 (1) (February 2008): 1-15.

[7] Pence, Gregory. *Medical Ethics*, 5th edition (Boston, MA.: McGraw Hill, 2008): 100.

[8] Ryan, Caroline. "More Than 3 Million Babies Born From IVF," *BBC News* (June 21, 2006): 1-3. http://news.bbc.co.uk/1/hi/health/5101684.stm.

[9] Medical Encyclopedia, "In-Vitro Fertilization," *MedlinePlus,* (January 31, 2007): 1-4. http://0-www.nlm.nih.gov.catalog.llu.edu/medlineplus/ency/article/007279.htm

[10] Scott, J.R. et al. *Danforth's Obstetrics and Gynecology, 9th edition*, (Philadelphia, PA.: Lippincott, Williams & Wilkins, 2003), ch. 39 "Assisted Reproductive Technology, p. 709.

[11] Centers for Disease Control, Department of Health and Human Services, *2003 Assisted Reproductive Technology Report,* (2004). http://www.cdc.gov/ART/ART2003/index.html. See 2003 National Summary Table.

[12] Collins, 3.

[13] Pence, 102.

[14] Tarkan, Laurie. "Lowering Odds of Multiple Births," *The New York Times* (February 19, 2008): 1-4. Http://www.nytimes.com/2008/02/19/healthy/19mult.html.

[15] Collins, 4.

[16] Collins, 4.

[17] Emory Reproductive Center, "Cryopreservation," Emory Health Care, 2008. http://www.emoryhealthcare.org/departments/ivf/services/cryo_preservation.html

[18] Collins, 5. Hoffman, D., Zellman, G., Fair, C. et al. "Cryopreserved Embryos in the United States and their Availability for Research," *Fertility and Sterility* 79 (5) (May 2003): 1063-1069.

[19] Embryo morphology (appearance of the cells/percentage of fragmentation) is one of the most influencial factors for embryo survival. Genetics & IVF Institute, "Human Embryo Cryopreservation (Embryo Freezing) and Frozen Embryo Transfer Cycles," (2008): 1-2. http://www.givf.com/fertility/embryofreezing.cfm

[20] Genetics & IVF Institute, 2.

[21] RAND/SART Working Group, 1.

[22] RAND/SART Working Group, 1-2.

[23] Nightlight Christian Adoptions, "Snowflakes Frozen Embryo Adoptions," (2008): 1-10. http://www.nightlight.org/snowflakefaqsgp.htm.

[24] Nightlight Christian Adoptions, 2.

[25] Keenan, J. "Medical Aspects of Embryo Donation and Adoption," Emerging Issues in Embryo Donation and Adoption Conference, Washington, D.C., (May 29-31, 2008): 1-16.

[26] Casey, S. "For Whose Sake?: The Current State of Human Embryo Donation and Adoption Law," Emerging Issues in Embryo Donation & Adoption Conference, Washington, D.C., (May 29-31,2008.): 1-12.

[27] Lysaught, 4.

[28] *Davis v. Davis*842 S.W. 2d 588, 592 (Tennessee, 1992)

[29] *Davis v. Davis,* 842 S.W. 2d at 597.

[30] Johnson, N. "Excess Embryos: Is Embryo Adoption a New Solution or a Temporary Fix?, *Brooklyn Law Review,* 68 (2003): 870-871.

[31] Weed, C. "Unscrambling the Uncertainty: Interpreting Egg Donor Agreements as They Relate to Embryo Adoption," Emerging Issues in Embryo Donation and Adoption Conference, Washington, D.C. (May 19-31, 2008): 1-16. Submitted for publication.

[32] Weed, 12.

[33] *In re Marriage of Buzzanaca,* 72 Cal. Rptr. 2d 280, 293 (California District Court of Appeals, 1998).

[34] Weed, 12. See also FLA. STAT. ANN. 642.11 (2) (2004).

[35] Weed, 13. See also 10 OKLA. STAT. 556 A. 1 (2004).

[36] Casey, 6.

[37] Kass, Leon. *Life, Liberty and the Defense of Dignity: The Challenge for Bioethics* (San Francisco: Encounter Books, 2002): 87.

[38] Moore, K, & Persaud, T.V.N. *The Developing Human: Clinically Oriented Embryology* (Philadelphia: Saunders/Elsevier, 2008): 15.

[39] Holy See, *Charter of the Rights of the Family,* no. 4: *L'Osservatore Romano,* November 25, 1983.

[40] Congregation for the Doctrine of the Faith, *Donum Vitae—Instruction on Respect for Human Life in its Origin and on the Dignity of Procreation—Replies to Certain Questions of*

the Day, (Rome: Congregation for the Doctrine of the Faith, February 22, 1987): 1-29. http://www.priestsfor life.org/magisterium/donumvitae.htm

[41] Brugger, E. C. "The Principle of Fairness and the Problem of Abandoned Embryos," Panel Discussion at the Conference on Emerging Issues in Embryo Donation and Adoption, (Washington, D.C.: May 30, 2008): 1-6.

[42] Lysaught, 3.

[43] Congregation for the Doctrine of the Faith, *Donum Vitae*, Section 1, No.5.

[44] Nonmoral evil refers to the lack of perfection in anything whatsoever. As pertaining to human actions, it is that aspect which we experience as regrettable, harmful, or detrimental to the full actualization of the well-being of persons and of their social relationships. For a more detailed description see, Louis Janssens, Ontic Evil And Moral Evil, in *Readings In Moral Theology, No. 1: Moral Norms And Catholic Tradition,* edited by Charles F. Curran and Richard A. McCormick, S.J. (Ramsey, N.J.: Paulist Press, 1979), 60.

[45] Richard A. McCormick, S.J., *How Brave A New World?: Dilemmas In Bioethics,* (Washington, D.C.: Georgetown University Press, 1981), 443.

[46] Richard A. McCormick, S.J. and Paul Ramsey, *Doing Evil To Achieve Good: Moral Conflict Situations*, (Lanham, MD.: University Press of America, 1985), 38. See also, Thomas Aquinas, *Summa Theologiae* II-II, q. 64, a. 7.

[47] According to McCormick and Ramsey, it can be argued that where a higher good is at stake and the only *means* to protect it is to choose to do a nonmoral evil, then the will remains properly disposed to the values constitutive of human good. The persons attitude or intentionality is good because he is making the best of a destructive and tragic situation. This is to say that the intentionality is good even when the person, reluctantly and regretfully to be sure, intends the nonmoral evil if a truly proportionate reason for such a choice is present. (Emphasis in the original) McCormick and Ramsey, 39.

[48] Lysaught, 3.

[49] Collins, 3.

[50] James J. Walter, Proportionate Reason And Its Three Levels Of Inquiry: Structuring The Ongoing Debate, *Louvain Studies* 10 (Spring, 1984): 32.

[51] McCormicks criteria for proportionate reason first appeared in Richard McCormick, *Ambiguity In Moral Choice* (Milwaukee, WI.: Marquette University Press, 1973). He later reworked the criteria in response to criticism. His revised criteria can be found in *Doing Evil To Achieve Good,* eds. Richard McCormick and Paul Ramsey (Chicago, IL.: Loyola University Press, 1978).

[52] Casey, 5.

http://dx.doi.org/10.5772/58592

[53] RAND/SART Survey, 2.

[54] IVF-Infertility.com, "Embryo Freezing: Results of Frozen-Thawed Embryo Transfer," IV-Infertility.com (2008): 1-2. http://www.ivf-infertility.com/ivf/frozen6.php.

[55] Genetics and IVF Institute, 1-2.

[56] American Fertility Society, "Ethical Considerations of the New Reproductive Technologies," *Fertility and Sterility,*Supplement 1, 46 (1986): 35S.